'An insightful and timely book that shows that "climate justice" is essential if we are to deal with climate change. The book forces us to acknowledge that those who have contributed least to greenhouse gas emissions are the most affected by the impacts of the climate crisis. This compelling book opens up the ethical and moral discussions of who is to blame and who should pay to deal with climate change.'

Professor Mark Maslin, author of *How To Save Our Planet*

'Why is climate change really about ethics and justice? Why isn't it just science, technology and politics? Cripps cuts to the heart of the matter, and in doing so risks breaking ours. Hauntingly clear, passionate, and unflinching. A powerful rebuke to those who want to deny the reality of climate injustice, and an uncomfortable, but stirring call to action for the rest of us.'

Professor Stephen Gardiner, author of *A Perfect Moral Storm: The Ethical Tragedy of Climate Change*

'It's about time someone gave us an accessible explanation of climate justice and why it must be integral to climate action. I thoroughly recommend Elizabeth Cripps' excellent book as a resource for students, activists, policy-makers and everyone else who wants to make a well-reasoned case for a just transition to a more sustainable world.'

Dr Sherilyn MacGregor, Reader in Environmental Politics, University of Manchester

'This book is a short and direct conversation with a philosopher carefully thinking through our duties now toward other people given the scary changes we all may face. While it may be painful reading at times, you will gain insights not available in any other book about climate change. The subtle analysis does not stifle the passion, and the deep feeling does not cloud the arguments. A moving philosophical plea for immediate radical action with the reasons distilled to their essence. If you wonder where to begin to tackle the worsening climate, start here.'

Professor Henry Shue, Merton College, Oxford.
Senior Research Fellow, Centre for International Studies, DPIR

'*What Climate Justice Means And Why We Should Care* is a compelling read and does a fantastic job at engaging a non-academic audience to envision the severity of the climate crisis. Through real-world illustrations and anecdotes, the book offers a concise overview of the complexities and moral dilemmas of addressing

climate injustice in the twenty-first century. Elizabeth Cripps stresses the importance of oppressions and discriminations based on social categories such as gender, race and class, and goes a step further to introduce the concept of multi-species justice which is often sidelined in discussions about climate justice. This book compels readers to think critically about injustice and the false dichotomy between individual and collective action.'

Pooja Kishinani, climate activist and co-author of *The Student Guide to the Climate Crisis*

'The iron law of global warming is: the less you did to cause it, the sooner and harder you suffer its effects. As this book makes clear, that raises very deep questions about justice, which we will be grappling with for the forseeable future. If you read this, you'll have a good head start on a crucial debate.'

Bill McKibben, author of *Falter: Has the Human Game Begun to Play Itself Out?*

'We live in a world increasingly impacted not only by climate change, but also its unjust impacts on both human and nonhuman communities. Elizabeth Cripps offers a lucid, comprehensive, and pertinent overview of a range of ideas and realities of climate justice in all its complexity. She offers the crucial argument that, in everyday political and personal practice, climate change is a *choice* to violate the rights of the most vulnerable. As inequitable as climate change can be, Cripps insists that it is possible, and straightforward, to choose climate justice instead.'

David Schlosberg, Professor of Environmental Politics and Director, Sydney Environment Institute, University of Sydney

'An essential primer. Elizabeth Cripps deftly explains the complexity of wicked problems without ever losing sight of the fundamental truth that, before it is a technical or political issue, climate injustice is a moral one.'

Professor David Farrier, author of *Footprints: In Search of Future Fossils*

'The concept of climate justice is increasingly being invoked. But what is climate justice? In her brilliant book, Elizabeth Cripps gives us a definitive answer. *What Climate Justice Means* shows why climate change is a matter of justice, who bears responsibility for this and what citizens and governments ought to do. It vividly conveys the realities of climate injustices and makes a compelling moral case for action.'

Simon Caney, Professor of Political Theory, University of Warwick

WHAT CLIMATE JUSTICE MEANS

And Why We Should Care

Elizabeth Cripps

BLOOMSBURY CONTINUUM
LONDON • OXFORD • NEW YORK • NEW DELHI • SYDNEY

BLOOMSBURY CONTINUUM
Bloomsbury Publishing Plc
50 Bedford Square, London, WC1B 3DP, UK
29 Earlsfort Terrace, Dublin 2, Ireland

BLOOMSBURY, BLOOMSBURY CONTINUUM and the Diana logo are trademarks of Bloomsbury Publishing Plc

First published in Great Britain 2022

For legal purposes the Acknowledgements on p. 215 constitute an extension of this copyright page

A catalogue record for this book is available from the British Library

Library of Congress Cataloguing-in-Publication data has been applied for

ISBN: TPB: 978-1-4729-9181-2; eBook: 978-1-4729-9182-9; ePDF: 978-1-4729-9183-6

2 4 6 8 10 9 7 5 3 1

Typeset by Deanta Global Publishing Services, Chennai, India
Printed and bound in Great Britain by CPI Group (UK) Ltd, Croydon CR0 4YY

For my family,
with love and hope

CONTENTS

Introduction

Imagine this. You and your partner and baby, rowing through foul water on an inflatable mattress you begged from a stranger. Your home and possessions submerged. Driven by fear from one refuge, where looters reigned and you saw a child die. You drift on your makeshift raft, past floating bodies, through the city that was once your home.

Or this. Another life, already precarious. A bamboo house, a village perched between jungle and sea. Another so-called 'act of God'. Your village is flooded, the salty water undrinkable. There was a fence between your community and wild animals, but the cyclone destroyed that too. You wait, knowing that a tiger could reach you and your family in five minutes. At any time.

Or this. Half a life, drought-wracked, scratched out on land made barren by heatwave after heatwave. Subsistence farming where there is no subsistence, where barely anything will grow and pests or disease strip what little there is. Escaping, if you can, only to another kind of poverty: violence, disease and choking pollution, on the fringes of a city already packed with desperate people. Or staying put, to starve.

For a moment, try to put yourself in these situations. Pick one. Close your eyes. Picture it. Your child, hanging on the edge of drowning. Your elderly mother, unable to run from a tiger. Your farmland arid, your family without water. Think of the choices you would face, the risks you could not avoid, the pain of hunger, of cold, of heat, of not being able to protect those you love the most.

Now step back. Be yourself again. But be yourself knowing that there are people living and dying like this, and there will be many, many more. Remind yourself that you are human and so are they. Ask yourself what that means.

For these are not a philosopher's thought experiments, plucked out of the air. The first is the story of a New Orleans delivery man, identified only as 'André', in the aftermath of Hurricane Katrina in 2005. He told it later, almost in passing, to journalist Jim Gabour. The second was reported by photojournalist Fabeha Monir, who travelled to southern Bangladesh in the wake of Cyclone Bulbul in 2019. The third is based on the future predicted by the Intergovernmental Panel on Climate Change

(IPCC), for countless African families. For many, it's already here.

This is what climate injustice looks like.

I'm not trying to make you feel guilty. Your guilt won't help, any more than mine will. But I want you to feel *something*. Because acknowledging the bleak reality of climate injustice is a first step towards recognizing our urgent responsibility to challenge it: to strive, instead, for justice.

We can no longer dismiss these extreme weather events – storms, floods, droughts – as natural disasters. Perhaps we cannot say with 100 per cent accuracy that climate change caused Hurricane Katrina or Cyclone Bulbul. But we know that events like these happen more often, and are more destructive, in our warming world, and that they will go on getting worse. We know that climate change causes disease and wildfires, decimates livelihoods. Of course, it's not your fault or my fault, individually, that this is happening. But it's happening because of the actions of governments of wealthy countries, multinational corporations and, in combination, people like you and me. Between us, we could change this.

So if someone asks me, 'Should we care about climate change?' I say that comes down to a simple enough question. Do we value those people's lives, or do we not?

Perhaps that sounds too simplistic. I don't think it is. I'm a philosopher, an ethicist. I've been writing, researching and teaching this subject for more than 16 years. Determining what *exactly* we should do is complex. Finding a balance between living your own life and fighting for a better world is tough. But recognizing the status quo as fundamentally terrible, knowing we have to act, and act now? That's uncontroversial morality.

Forget political side-taking. Stripped of the polarization driven by vested interests, climate justice is about what it means to acknowledge our fellow humans as just that: human, like you and me. It's about the wrongness of violating human rights, of causing malaria and starvation and fires that choke families to death in their own homes. It's about recognizing that we should save our fellow humans from terror and suffering, if we can. It's about the baseline for a minimally just world, and how far we have fallen below it.

This book starts there. It turns a mirror on the way we live now: institutionalized dependence on fossil fuels, systematic disregard for many human lives, radical discrepancy between those who have a say and those who feel the pain. It rejects the reasons we might find for not acting, for thinking it's not our problem. '*I'm* not harming anyone.' 'We don't owe justice to distant strangers, let alone people who aren't born yet.' 'It's the economy, stupid!' 'It's too hard.'

Climate injustice *is* racial injustice, *is* gender injustice, and is unjust in a way that cannot be reduced to either, because the two intersect. I'll talk about why. I'll discuss the legacy of colonialism and entrenched racism and sexism. I'll highlight failure of recognition and denial of political voice. I'll ask, too, if what we are doing is an injustice to non-humans: to other animals, species, ecosystems. The starving polar bear, the devastated Amazon. Of course, it matters *for us* that we are destroying the natural world. We need it. But would this be wrong, even if we didn't?

Then I'll look forward, to what climate justice would look like. Mitigation, or limiting global warming to a 1.5°C rise. Above that, the IPCC warns, things will get dramatically worse. Adaptation, or changing systems and developing technology, so that everyone can live with that 1.5°C. Compensation, where these come too late. And, crucially, listening to the dispossessed, the victims of climate change, in deciding *how* to do all this.

But who should pay, in a world where no one wants to? States, firms and individuals with the biggest carbon footprint, because they're causing harm? The very rich, because they can, or because they're free riding on the pollution of generations? And where do we go if not towards true justice? Because, in practice, international negotiations come nowhere near. What should each of us do, as individuals? Change how we think and how we live? Become activists?

Both.

In September 2020, in the middle of a global pandemic, young people refused to forget this still greater emergency. Climate protests took place in 3,500 towns and cities around the world, under the hashtag #FightClimateInjustice. *They* get it. I won't say they give us hope, because it's not their job to do that. Leaving children to solve this escalating crisis is, itself, a dereliction of responsibility. Instead, I'll show why others must follow their lead.

Some think there is no time for ethics: that it is a distraction in the all-consuming urgency of 'fixing' climate change. This book is about why they are wrong.

Climate change is not a purely technical challenge. Scientific innovation is essential, but we cannot develop the right technology, or use it, without knowing what we want it to achieve. Political solutions are crucial, but we need something to judge political compromises against. We need moral arguments.

Any climate policy makes assumptions: about what takes priority, who to protect, whose voice to take seriously. Philosophers expose those assumptions, so they can be judged. Skip this step, and it doesn't mean we have a value-free way of making decisions. It means the 'values' doing the work, too often, are the interests of those in charge.

As a moral philosopher, I will use clear reasoning from fixed points: facts, strong intuitions, principles so generally accepted and philosophically grounded that they can be taken as fundamental. I'll draw my own conclusions about these desperately important questions. But I also want to give you a philosopher's toolbox, to do the same.

Of course, the moral story cannot be told in a vacuum. I won't pretend it can. We need science and social science, because without knowing where we start and what we *can* do, we cannot say what we should do. Our own perspective matters, too. I'm a philosopher, trained to apply rational thought to real-world dilemmas. But I'm also an activist, a mother, a *person*, grappling with these challenges for myself.

And I'm a white woman with a good job in a rich country, who knows that my own privilege could blind me. You, too, may be biased by advantages you have always taken for granted.

Many of my examples are *not* hypothetical, because they remind us of what is at stake: no abstract tangle of ideas but a real crisis, with millions of lives in jeopardy. Some passages may be hard to read, but I make no apology. They have been hard to write. If I don't pull any punches, it's because it is too late for that.

There is room for hope, even in this terror. But there is no room for passivity.

I

Basic Justice, Incontrovertible Science

Let's start with a simple premise. Humans have value. It matters that men, women and children can lead good rather than terrible lives. It matters that we don't seriously harm our fellow humans, that we protect them if we can.

We *know* this. Mostly, we know it instinctively. We think of Dr Harold Shipman, who murdered his patients for gain. We judge him. We read of the sadistic cruelty of slave-owners, making exploitable objects of human lives. We know this for what it was, and is: unutterably wrong. 'Are there no prisons?' demands Ebenezer Scrooge, the rich miser in Charles Dickens' *A Christmas Carol*, refusing to help the destitute. 'Are there no Union workhouses?' We have no problem judging him, either.

Philosophers fill this out. So let's do the same.

UNCONTROVERSIAL MORALITY

Morality is about how we ought to live. There are things you cannot do to other people without denying either their moral status as humans, or that you are yourself a moral agent. That is, that you are someone to whom moral rules apply, capable of understanding and following them. There are also some things you *must do* for your fellow humans if you are to recognize them – and yourself – as such.

Take John Stuart Mill's No-Harm Principle. It is wrong to do serious, foreseeable, avoidable harm to another human being. To maim them, torture them, starve them, fill their homes with water. If ever there was a clear-cut moral baseline, this is it. Some put it another way: we mustn't violate human rights.

This is a cornerstone of criminal and international human rights law, as well as what philosophers call 'common sense morality': the rules we use to judge each other, and ourselves, in day-to-day life. It's a basic tenet of liberal thought, whether you get to your principles teleologically (start with the importance of human lives going well and figure out what set of obligations would best preserve that) or deontologically (start with what it means for the duty-bearer to acknowledge the moral status of their fellow humans). But it transcends Western ways of thinking. For example, *Ahimsa*, or non-harming, is a precept of Buddhism.

We can go further. Responsibility matters, but it's not the only thing that does. To recognize someone as your fellow human is to recognize the claim that their need has on you, even if you didn't cause it. Being able to help is, itself, a moral reason for helping. The philosopher Peter Singer ties these together: if you can save someone from severe suffering, comparatively easily, you should.

Few would deny this in individual cases: the heart attack victim you could save with a call to the rescue services, the child you could pull back from the cliff-edge. Collectively, it's a scarcely less controversial claim, also recognized in international law. Everyone is entitled to some level of human flourishing, the opportunity for a *decent life*. To put it another way, with human rights come not only 'negative' duties (not to violate others' rights, not to torture, not to kill) but also 'positive' ones (to protect them, to provide healthcare and nutrition).

In fact, we don't have rights at all, in any meaningful sense, unless they come with both kinds of duty. The philosopher Henry Shue pointed this out four decades ago and it's still true now, however adept politicians are at ignoring it. What does it mean to have a right to physical security, if there's no infrastructure to uphold it? If you're at the mercy of any mugger or abuser who chooses not to recognize your common humanity? Your life starts to look like those in Thomas Hobbes' dystopian state of nature: 'nasty, brutish and short'.

If morality is about how we ought to live, justice is about how we must live *together*: how we should organize our societies and institutions. Collectively, then, this piece of fundamental morality – this claim each of us has to a decent human life – becomes the baseline for a minimally just society.

But how far does this go? Having established this moral boundary, we have to understand what it actually means. What *is* 'serious harm'? What is a 'decent life'? What do we have this priority-taking, universal moral right *to*?

WHAT WE NEED TO LIVE

For our lives to go well – for them to go at all – we need some core things. Health. Nourishment. Shelter. Ownership of our own bodies. Most accounts go further than this. A 'decent life' is a higher bar than 'a life worth living', which only amounts to 'better than no life at all'.

For a decent life, we need community, education, a political voice. We need adequate options and the ability to choose between them. Many think we need more. The philosopher Martha Nussbaum combines Aristotelian philosophy with insights from Indian women living in poverty. The result is a list of central human interests, or 'capabilities', which she thinks people from different cultures could agree on. For this more nuanced 'flourishing', we need 'Emotions', including the chance to form loving connections, not to be blighted by trauma or abuse. We need 'Senses, Imagination and Thought', which require freedom of expression and religion, and 'being able to search for the ultimate meaning of life in one's own way'.

Another capabilities model, that of the development economist Amartya Sen, underlies the UN Sustainable Development Goals: no poverty, zero hunger, good health and well-being, quality education, clean water and sanitation, decent work. The 1948 Universal Declaration on Human Rights includes rights to political participation, freedom of expression and religion, education, and 'a standard of living adequate for the health and well-being of self and family, including food, clothing, housing and medical care and necessary social services'. Human need is already formally recognized, if, in practice, inadequately protected.

Controversy remains around the edges. Do we have a central interest in being able 'to laugh, to play, to enjoy recreational activities'? Nussbaum thinks so. Certainly, children need play for well-being and development. But is it indispensable for a decent adult life?

And do we need to form ties with non-humans? Again, Nussbaum says we do. My intuition is that she's right. But is she? Of course, we depend on the so-called natural world for our other central interests. This book will make that clear. Respectful connections with places, ecosystems and other species are also central to many non-Western or Indigenous ways of life. Even in the high-income countries collectively known as the 'global north', many of us value getting out into the forest or hills, watching birds dip and dive on a wild coast. Non-humans are beloved companions. In London, the World Happiness Report found that being in parks and allotments, around ponds, lakes, canals and rivers, made people happier. But can no one lead a full life without such active interactions, *even if they think they're perfectly happy*? That's harder to prove.

It matters that we get this right. It matters for what we should aim for in our societies, for what 'full' justice requires (whatever that is). It matters because we should think not only about avoiding climate *in*justice, but also more positively, about the kind of world we want to live in together. There's psychological evidence that the ceaseless pursuit of materialism, stepping away from nature, makes us *un*happier.

But it doesn't matter for getting this book off the ground. We don't need an expansive idea of a 'good' life to know that the climate crisis constitutes a massive injustice. Climate change undermines indisputably basic needs. It kills, it inflicts disease, it rips away livelihoods. It harms.

REAL CHANGE, REAL HARM

This is already happening. 'Climate change', says the 2021 report from the IPCC, 'is already affecting every inhabited region across the globe.' It is shrinking glaciers, thawing permafrost and reducing crops. It's worsening heatwaves, droughts, floods, cyclones and wildfires. It's exacerbating poverty, killing with heat and cold.

Even in rich countries, people are losing everything. More than 1,500 people died from Hurricane Katrina in 2005. In winter 2021, hundreds died in storms in Texas. Less than six months later, wildfires raged across Greece, Turkey and Italy, killing several people and displacing thousands. More than a hundred died that same summer in flash floods in Western Europe, dozens in Australian bushfires in the summer of 2019/20 and in Californian wildfires in 2020. Thousands of homes were destroyed. In 2021, the anthropologist Juan Francisco Salazar and colleagues interviewed more than 6,000 Sydney residents. All said the bushfires had affected their sense of well-being.

In the global south, this has been the reality for decades. Each year from 2008 to 2016 an average of 21.8 million people were internally displaced by extreme weather. In the first nine months of 2017, some 3.2 million people in low-income countries were forced from their homes by these so-called natural disasters. (The figures are from Oxfam.) Behind each of the headline incidents – Cyclone Winston (2016) displacing 55,000 people in Fiji, 1,200 killed in a month by floods in Bangladesh, Nepal and India – lie countless individual losses. Myriad stories of human pain, each one a tragedy.

Think about just a few of the vulnerable children of Bangladesh, as a UNICEF report does. Eleven-year-old Maroof Hussein, his school and house flooded;

his eight-year-old friend, Iqbal, drowned. Fourteen-year-old Nazma Khatum, her dreams of nursing crumbling, at risk of child marriage. Mohamed Chotel, 13, collecting discarded plastic in a Dhaka slum. Sonia, 15, driven by river erosion to the city's filthy underbelly, by rape to prostitution.

This is what climate injustice looks like. And it's only the beginning, if we don't take collective action.

The IPCC catalogues risks. Water restrictions, more flooding, more deaths from heatwave and wildfires. That's a hefty economic toll. It's a vast human one too. In the UK, deaths from heatwaves will increase 250 per cent by mid-century, on predictions from the Priestley International Centre for Climate. Within two decades, increased flooding could put 2.6 million people at 'significant risk'.

Think about this, but not as a detached observer. Think about it in relation to people you care about. Think about it as a parent, if you are one. Today's children face the prospect of sea salt in drinking water; soil quality and crop yields threatened. They face ever-increasing chances of malaria, dengue fever, encephalitis.

For those who are already vulnerable, things will get much, much worse. The World Bank economists

Julie Rozenberg and Stéphane Hallegatte believe that due to climate change a further 122 million people could be living in poverty by 2030. Countless lives ruined by malnutrition, by disease. Crop failure, water stress, the encroaching seas, would turn land from productive to barren, from home to unliveable.

Two hundred million people could be displaced by climate change. So said Norman Myers, trailblazing environmental scientist, in 2001. The World Bank predicts 86 million internal climate migrants in Sub-Saharan Africa by 2050, 40 million in South Asia, 17 million in Latin America. Envisage it – for each of those individual men, women and children, each with their own interests and plans for the future, their favourite foods and favourite jokes – before they're dehumanized, in your mind, by the scale of the tragedy.

In 2014, Enele Sopoaga, then prime minister of the nine small islands of Tuvalu in Polynesia, implored his fellow leaders: 'I ask you all to think what it is like to be in my shoes. If you were faced with the threat of the disappearance of your nation, what would you do?' For Tuvalu, for Kiribati, the Marshall Islands and the Maldives, climate change jeopardizes their physical survival.

This is the future we are bequeathing. A life without the security of being able to make plans, without knowing whether you can go on breathing, eating, feeding and educating your children, because everything could be ripped away at any moment. Factor in the possibility of abrupt, irreversible change – large-scale ice sheet loss, pushing up sea levels – and, if you're not worried, you haven't been reading carefully. The higher temperatures go, the higher we *let* them go, the greater the suffering.

The future, in fact, is brutal.

A MENTAL HEALTH CRISIS

A New Orleans resident Brandi Wagner got through Hurricane Katrina, but her mental health didn't. Twelve years later, she told *Politico* what her life had become: a battle with depression, anxiety and opioid addiction, unable to work.

Trauma, shock, post-traumatic stress disorder, compounded stress, strained social relationships. Depression, anxiety, substance abuse, aggression and violence. Suicide. Loss of autonomy, control, place, personal and occupational identity. Helplessness, fear, fatalism. *Solastalgia*, or grief for a changed environment. It's a grim tally, but it's what the

American Psychological Association thinks we should expect from climate change.

'We come back from our field sessions increasingly broken,' marine biologist Steve Simpson told the *Guardian* in 2020. In 2013, psychologist John Fraser and colleagues surveyed 182 conservation biologists, environmentalists and conservation educators. They're the ones who should know, and 63 described themselves as constantly worried about the current state of the planet; 89 were constantly worried about future environmental conditions.

Climate change is, quite literally, a nightmare for children. Of more than 2,000 aged between 8 and 16 surveyed for the BBC's *Newsround* in 2020, 19 per cent had had a bad dream about it. Almost three-quarters were worried about the state of the planet and 22 per cent were very worried; 58 per cent were worried about the impact climate change would have on their lives. The same year, more than half of child and adolescent psychiatrists in England had patients distressed about the climate emergency, according to figures from the Royal College of Psychiatrists.

And that's in the UK where, for the majority, this is still a future tragedy: a stifling fear. But for Brandi Wagner in Louisiana, for the many already tormented by climate change in poorer countries, the trauma is now. In 2021, psychotherapists surveyed 10,000 young people from around the world: 84 per cent were worried about climate change, and 59 per cent very or extremely worried. In 2020, a child from the Maldives in the Indian subcontinent told researcher Caroline Hickman: 'Climate change is like Thanos in *Avengers: Endgame*, whose ideology is to kill off half the planet so the other half can thrive; trouble is, we (in the Maldives) are the half being killed off.'

'IT'S NOT THAT SIMPLE!' (ACTUALLY, IT IS.)

So far, so painfully straightforward. Climate change is an injustice in the most basic sense: a violation of human rights, a morally impermissible harm. It is also a failure to protect the most vulnerable. But, still, there is inaction. So let's turn to the excuses we might make for not taking this seriously, and see whether they hold up.

Excuse 1: 'The science is still up for debate'

The science is not in doubt.

The IPCC confirmed the existence of anthropogenic (human-caused) global warming in 1990, though there was plenty of evidence before then. Ever since, it has become more certain and more stark. The 2018 IPCC report brought an urgent message: there's a narrow window to stave off catastrophe. (Narrow enough then; rapidly closing now.) Things will be bad enough if temperatures reach 1.5°C above pre-industrial levels. They'll get much worse with a 2°C rise; still more disastrous with a further rise. Keeping below 1.5°C means cutting carbon dioxide emissions by 45 per cent from 2010 levels by 2030, hitting net zero by 2050. On the 2021 IPCC figures, meeting that target is still possible, but not without urgent action. By 2019, global surface temperatures were already around 1°C hotter.

In 2004, Naomi Oreskes, history of science professor, checked 928 peer-reviewed articles on climate change, from 1993 to 2003. None denied the link between increased greenhouse gas emissions and rising temperatures. In 2009, Peter Doran and Maggie Zimmerman surveyed 3,146 Earth scientists. Ninety per cent accepted climate change as real; 82 per cent agreed that human activity had significantly contributed to causing it. Of the climate specialists, *97 per cent* accepted anthropogenic climate change.

And yet, exposed to large sections of the media, listening to some policymakers, you could easily believe the opposite. You'd think a few alarmist scientists were spouting made-up warnings to scare us away from this manna from heaven, this source of all jobs: oil, coal and gas.

There's a reason for that, and it's not that the scientists are wrong. Doubt has been carefully manufactured. The evidence on this from the social sciences is as frightening as the climate science it defends. It shows that climate denial is an organized industry, born in the US but now spread across the world.

Money talks. A cliché, because it's true. It talks to politicians. In 2017, *Inside Climate News* reported that ten oil refiners poured nearly $30 million into opposing a carbon fee in Washington State, including $11.6 million from BP America. (The biggest donation in support of the fee was $3.4 million from the Nature Conservancy; the total from the top ten supporters was $9.45 million.) In 2019, Fossil Free Politics, an amalgam of NGOs, announced that BP, Chevron, ExxonMobil, Shell and Total spent €123 million on lobbying the EU between 2010 and 2018. Their Brussels-based lobby groups spent another €128 million.

Here's something else the prime minister of Tuvalu told his fellow heads of state: 'It is the eyes of the children that we must answer to, not the fossil fuel industry.' Too often, governments do exactly the reverse.

And it's not just fossil fuel companies. Nine of the ten biggest US meat and dairy firms have spent $109 million between them on lobbying and $26 million on political campaigns since 2000, according to 2021 research by Harvard PhD student Oliver Lazarus and colleagues. (The other giant, Koch Foods, was excluded only because there was no publicly available information on its political activities.) Eight of the ten directly lobbied Congress and the Environmental Protection Agency.

Money talks to the public. The sociologist Robert Brulle and colleagues assessed the advertising spending of five major oil and gas companies between 1986 and 2015. The more the media covered climate change, or the US Congress paid attention to it, the more fossil fuel advertising dollars went up. A year later, Media Matters research made the link clearer still. In the two one-week periods after it was announced that 2015 was the hottest year on record and February 2016 the most abnormally hot month, CNN viewers saw nearly five times more American Petroleum Institute advertising than they did coverage of either climate change or the temperature records.

Oreskes and her colleague Geoffrey Supran examined internal documents, advertorials (paid newspaper content) and peer-reviewed research by ExxonMobil employees between 1977 and 2014. According to the peer-reviewed papers and internal reports, climate change is serious, anthropogenic and solvable. The advertorials, the social scientists said, sow doubt about all three. Their conclusion? That the fossil fuel giant misled the public.

Money talks to children. In 2017, a conservative thinktank circulated a book to hundreds of thousands of American science teachers. It was called *Why Scientists Disagree About Global Warming*, credited to the carefully named *Non*governmental International Panel on Climate Change. According to the Center for Public Integrity, the Oklahoma Energy Resources Board has funded pro-fossil-fuel educational materials: curriculums, speakers, after-school programmes. The highlight? A series of books about a cartoon character, 'Petro Pete'. Pete thrives on oil products, but occasionally faces the nightmare that they might be snatched away, leaving his life empty. (I wish I were making this up, but I'm really not.)

Perhaps most depressingly of all, money talks to an influential handful of scientists. The ten biggest US meat and dairy firms have all, say Lazarus and colleagues, 'contributed to research that minimized the link between animal agriculture and climate change'. As for the fossil fuel industry, read *Merchants of Doubt*, the 2010 exposé by Oreskes and Eric Conway. But only if you've got a strong stomach.

It tells the story of a few high-profile scientists, used to media attention, close to politicians, commenting on areas generally outside their own expertise and, in the process, destroying the lives and reputations of peer-reviewed experts. This is a tale of deliberate undermining of scientific consensus on a host of issues: the link between tobacco and smoking, acid rain, the hole in the ozone layer, the harm done by pesticides, and – of course – climate change. It's the story of doubt as a (very lucrative) product, of cherry-picking data, of magnifying yet-to-be-explained details, of exploiting general ignorance of how science works.

The result? A lingering impression of ongoing debate within the scientific community, completely at odds with what is demonstrably true. A cover-up that might just kill us all.

The science is right, despite all these attempts to undermine it. So let's look, instead, at where *moral* doubt might creep in.

Excuse 2: 'We don't owe justice to distant strangers . . .'
Here's another seemingly uncontroversial idea. We owe more to the people closest to us. On one version of this, each of us has obligations to the people we share a state with, that we don't have to people in far-flung parts of the world.

This goes back to a core liberal notion: the social contract, which found its dominant modern expression in the work of philosopher John Rawls. Think of the state as a cooperative scheme for mutual advantage, collectively enforced. We're all playing our part; we all accept some curbs on our natural liberties and in return we get the protection of justice. We each owe it to one another to follow the rules of this mutually beneficial game: obey laws, pay taxes, do our 'bit' in making sure everyone has an equal opportunity to do well. In other words, justice amounts to institutionalized reciprocity. For some philosophers, so-called 'statists' or 'nationalists', this reasoning allows us to draw a clear line: what we owe our fellow citizens versus what we owe the rest of the world.

Examine this more deeply, though, and it doesn't look so clear. We can't now think of ourselves as living in mutually isolated cooperative bubbles (if we ever could). We are linked by global trade and international law, not to mention climate change. Think about a Kenyan farmer, growing coffee that will eventually be sold in London, Paris and New York. Her entire livelihood, whether she can eat or drink or educate her children, depends on global prices. She's as coerced by one 'shared institution' – the web of economic ties binding her to the London café-goer downing his espresso – as is the Scottish farmer who grew the oats for that Londoner's porridge. The only difference? She *doesn't* get to enjoy any security in return, or get to vote on how the system functions.

There's also a more fundamental question: why should justice apply only within certain institutions? Why should it hang or fall on mutual advantage?

In asking this, we come back to what 'basic justice' actually means. This chapter hasn't been talking about equality, beyond our equal moral status *as* humans. *Full* justice – social or global – might well require equality of opportunity, or some version of it. Plenty of political philosophers think so. But you don't need to agree, to see why climate change is fundamentally unjust. This is about people being able to lead decent lives: not drowning, not dying of malaria or malnutrition.

Really protecting people can't be reduced to distributive justice, or 'who gets what'. It is about who has a voice in decision-making, who is seen and valued. And it is about not doing serious harm. The philosopher Simon Caney pares the argument to its moral bones. Take the human rights not to be killed, not to be deprived of health or of the means of feeding yourself. Climate change, caused by humans, suffered by humans, violates these.

Let's be clear. I'm not saying *in*equality never matters, even for basic justice. Severe or systematic inequalities – race-based, gender-based, class-based – undermine self-respect, perpetuate oppression. When money buys influence and only a minority *have* the money, they have an unpleasant habit of treating everyone else as dispensable. We've seen that already, in climate change denial. We'll see it again.

But for now, what matters is this. To see climate injustice for what it is we need only recognize our shared status as human beings, and what it means to be a moral agent, at all.

Excuse 3: '. . . or to future generations'

What about the victims of climate change who haven't yet been born, who will contribute to our societies only after we are dead? Put bluntly, Excuse 3 goes like this: 'They can't do anything for us, so we needn't do anything for them.'

This is a flawed argument. It's flawed even if you buy the idea of justice as a cooperative scheme for mutual advantage. Future generations may not be able to pay us for any benefits we leave them (except by protecting our reputations or preserving our projects). But we, and they, can reciprocate *indirectly*, across generations. We can repay benefits from past generations by handing them down to future ones. The ethicist Lawrence Becker calls this a stewardship model. Alternatively, we might owe it to those who share our societies *now* to leave their descendants the opportunities that we (and they) have enjoyed. That's how Rawls himself makes sense of it.

Like citizens of other states, future generations are effectively coerced by current political and economic structures. They just don't get to be part of the decision-making process. Once again, it seems exceptionally unfair to exclude them from the entitlements of justice. Or think of it another way. Future generations are made up of people whose very existence is our responsibility. There is something peculiarly abhorrent in saying we owe them less than we owe each other, because we can't get anything from them in return.

Moreover, we don't have to buy the Rawlsian model. Once again, we're talking about *basic* injustice. Harm doesn't become acceptable just because the person you maim, kill or starve is in a different generation. Suppose, says Henry Shue, you decide it would be fun to plant landmines on a popular tourist trail. It's morally abominable if they go off tomorrow; it's still abominable if they go off in 150 years' time. Even if you didn't plant them, you should remove the landmines if you easily can.

Excuse 4: 'We're not harming future people'

It is wrong to harm future people. But *are* we harming them, by causing climate change? Our next excuse denies that, despite the scientific evidence.

Version one goes like this. We don't *know* we're harming future people because we don't know what they'll be like, what they'll care about or want to do. Why should they miss the rainforest, polar bears, or beaches lost under rising seas, if they never knew them? They might flourish in glass bubbles, not a tree in sight, fed by tubes, walking an hour a day on a treadmill, living their dreams through sophisticated computer games.

Well, for a start, we might think there's something inauthentic, unhuman, in a life whose joys and sorrows are entirely artificial. (I do, for one.) But there is, again, a more basic point. Climate change doesn't just undermine luxury life choices or upset those dismissed as 'tree-huggers', who happen to care about the rest of the planet. It violates human rights. It strips people of the ability to lead *any* kind of decent life.

Even if our descendants don't care about the rainforest, they will need us to protect it. People in a hundred or a thousand years' time will need to live, absorb nutrients, be healthy and free from pain, just as we do and our ancestors did. There is no reason to think future generations will be indifferent to being drowned, burnt, mutilated, starved, exposed to disease, or poisoned by toxins from plastic waste. They'll plausibly need a great deal more than that: education, community, scope for autonomy. But even if they don't, the point remains. Climate change will make life bad, in many ways, for future people.

There's another version of the excuse. It accepts what I've just said, but still denies that climate change harms anyone born in the future. If you're not interested in abstract puzzles, you can skip it, because it's not something you'll catch anyone but a philosopher worrying about.

The argument goes like this. We don't harm someone if, in causing them to suffer, we *also* cause them to exist in the first place (at least, so long as their life is still worth living). Pick a future person, born in 2080 to conflict, escalating poverty, searing heatwaves and rampant disease. Call her Penny.

Penny is badly off. But in failing to cut greenhouse gases, this generation didn't make her worse off than she would have otherwise been. If we had taken decisive action on climate change, Penny would not have had a life at all. If we had chivvied institutions, set binding targets, transformed the power and transport industries, our lives would all have been different. People would have had different jobs, different ways of getting about. They would have met different partners or met them at different times, had sex on different days, and conceived different children. Penny's great-grandparents might have had kids – together, or with different partners – but they wouldn't have had her grandparents. Penny would never have existed.

As a philosophical puzzle, this is intriguing. It even has a name, the non-identity problem, because the challenge is to explain *why* it's wrong to create a person whose life would be bad in many ways, rather than someone else, with a different identity, whose life would be better. It's prompted reams of academic writing. But as a reason not to care about climate justice, it's unconvincing. Even Derek Parfit, the philosopher who came up with it, was confident there was a way round.

Think what it would mean to take the argument seriously. We could do anything we liked to future people, so long as in the process we changed which individuals were born and didn't leave them with lives not worth living. (A lower bar, remember, than denying them a decent life.) We could deliberately conceive severely disabled children, so we could study them. We could pour chemicals into waterways, poisoning thousands in three generations' time, so long as the chemicals radically changed patterns of attraction now, so that different children were born.

The non-identity problem is no moral 'get out of jail free' card. Instead, it should (and does) prompt deeper reflection on what exactly we owe future people. We might think in terms of group interests or rights. Or of what we owe not to individuals, but to whoever stands in a particular relation to us, as the philosopher Casper Hare suggests. There are things I must do for my daughters not because they are them, those much-loved individuals, but because they are *my children*. Think like that, and it's not a question of what I, or this government, or this generation, owes Penny. It's about what we owe to whoever happens to be born into that generation.

Perhaps it's possible to harm or wrong someone without making them worse off than they would have been otherwise. Depriving someone of a central interest usually makes them worse off overall. But wouldn't it be wrong *even if it didn't*? Seana Shiffrin imagines a wealthy eccentric giving someone a million dollars by dropping it from the sky in a cube. The cube breaks their arm. Another philosopher, James Woodward, describes a racist airline official who won't sell a ticket to a man of colour. The plane crashes with no survivors. Both these people have made someone better off overall, but they've still behaved wrongly towards them.

We can unpick the non-identity problem. But even if we couldn't, it only works if future people have lives worth living. Many lives will be made so bad by climate change – their suffering so unendurable – that they would rather not have lived at all. That's what the group of young women known as 'BirthStrikers' highlight: they won't have kids because they don't want to bring them into this damaged world.

Nor is climate injustice 'only' a matter of harm done. It's also about the minimum we *owe* others in a positive sense. Basic rights, basic interests, the chance of a decent life. That applies whatever generation they are born into, whoever they happen to be.

Excuse 5: 'Not my fault, not my job'

So we admit the harm. Climate change destroys lives and destroying lives is wrong. The next piece of 'Yes, but . . .' reasoning goes like this.

Who is doing this? Whose job is it to prevent it? As individuals, our impact seems infinitesimal. Even a rich state can't stop global heating alone. There's no single global institution responsible either for causing climate change or for protecting people from it. It's tempting to retreat into the security of apparent powerlessness: 'I don't make a difference.'

In other words, this may be *a* problem, but it's not *my* problem.

We're used to thinking about responsibility for individual actions ('Joe broke Jim's nose'), or about collective responsibility, when a group of people do something intentionally together ('Joe, Jim and Alesha trashed the bird hide', 'that corporation misled the public'). However, we can also share responsibility, or be collectively responsible in a weaker sense.

We blame people, or hold them responsible for harm, when they knew or should have known that their actions would hurt someone. With climate change, we each know that our combined acts – acts we want to do and know others will want to do – cause desperate suffering. It's a plausible extension of standard thinking about responsibility, to say that 'we' (together) are responsible for the ensuing harm, and should prevent it between us.

Or think about it the way climate ethicist Stephen Gardiner does. As individuals, we're part of patterns of behaviour. Some of us are fliers, many are drivers. Many are shareholders, especially in affluent states, often through pension funds. We are consumers of oil, gas and coal, dairy and meat. In the language of another philosopher, Christopher Kutz, most of us are complicit.

Looking back, we judge past generations for just such complicity. We recollect the abominations of the Nazis or the slave trade and we know them for what they were: monstrous. It is, of course, infinitely harder to turn this critical lens inwards, to see the way of life dominant in the global north for what *it* is: a consistent violation of others' most basic rights. But that doesn't make it any less true.

Still unconvinced? Flip the discussion around. Consider this: a true story. The year is 2013, in Essex, in the UK. An elderly man's car skids off a bridge. He's disabled, trapped by his seatbelt, sinking with his Lexus. Three passers-by jump into the river. They work together, push the car into shallower water, keep him afloat until the emergency services arrive.

Set aside complicity for a moment, and remember the other moral principle we started with. We should save our fellow humans from harm, even if we didn't cause it. Especially if it's easy for us. Sometimes we can't do this alone but we can by cooperating. 'We're not heroes,' Spencer Turner, one of the Essex rescuers, told the press afterwards. 'We just did what anyone would do in that situation.' That may or may not be true, but this is. They did what any three fit, able-bodied adults *should* have done.

Scaled up, fulfilling that basic duty means at once much more and much less than being constantly on the alert to perform *ad hoc* rescues, cooperating with friends or strangers. As Henry Shue points out, consistently protecting our fellow humans means building structures and institutions to keep everyone safe.

There's a temptingly easy response: 'We *have* institutions. This is the government's job!' But it doesn't let us off the hook. Yes, governments should cut greenhouse gases, invest in technology, fund infrastructure to allow societies to adapt to some global warming. However, our governments act on our behalf. If they don't enable us to fulfil our moral obligation to others, it falls back on us. What's more, climate change is a global, intergenerational challenge. Remedying climate injustice requires cooperation beyond state borders. That means looking beyond governments.

This isn't about blaming individuals. That would be an injustice in itself. But climate justice is about having institutions fit for purpose: institutions to fulfil the shared obligations that we all have, as humans, to protect one another's most vital interests. That's not where we are now.

2

The Same Storm, But Not The Same Boat

On 24 January 2020, a news agency circulated a picture of climate activists at the Davos World Economic Forum. It was an image for the moment: in the background, snow-tipped mountains, a climate bellwether; in the foreground, four resolute young women, challenging injustice. There was just one problem. It wasn't the whole picture. The four faces were white. Vanessa Nakate, a Black Ugandan activist, had been cut out.

The agency later apologized, calling the crop an honest mistake. But the decision, error or not, was too unpleasantly symbolic to ignore. Women of colour face the greatest threats from a climate emergency precipitated, for the larger part, by the decisions of white men. Non-white and female voices often go unheard. Now, with the click of a mouse, this eloquent activist, with justice on her side and insights her European contemporaries could never share, had been erased. As she told the *Guardian*, afterwards, 'It was like I wasn't there.'

Nakate's experience was an insult to her, and to the generations of women of colour fighting for environmental justice. But it was not surprising. If the voices of Black women and girls had been heard – if their needs had been factored into centuries of global decision-making – we would not be in the situation we are now.

DRIVING CLIMATE CHANGE

Climate change is about privilege. Causing it, and not (yet) worrying about it. In 2018, just ten countries, listed in Table 1, accounted for around 60 per cent of global emissions.

But headline statistics can be misleading: some countries have a lot more people to sustain than others. The UK didn't even make the chart, with its 441.13 tonnes of CO_2 equivalent. But average per capita calculations tell a different story. The UK produced 6.64 tonnes per citizen: well over twice that of the average Indian. In Ghana, emissions per person were only 0.66 tonnes; in Fiji they were just 0.32 tonnes.

TABLE 1: *Greenhouse gas emissions 2018 (metric tonnes CO_2 equivalent)*

Country	*Total*	*Per capita*
China	11705.81	8.4
United States	5794.35	17.74
India	3346.63	2.47
Russia	1992.08	13.79
Indonesia	1703.86	6.37
Brazil	1420.58	6.78
Japan	1154.72	9.13
Iran	828.34	10.13
Germany	776.61	9.37
Canada	763.44	20.40

Source: Climate Watch (World Resources Institute)

According to Oxfam and the Stockholm Environment Institute, the richest 10 per cent of people in the world accounted for 52 per cent of carbon emissions from 1990 to 2015. The richest 1 per cent produced more than twice as much as the poorest half of the world. On Climate Accountability Institute figures, the 20 biggest oil, coal and gas companies produced 35 per cent of worldwide emissions between 1965 and 2018.

Income and wealth may sound like neutral categories when it comes to climate, but they're anything but.

In 2020, Oxfam reported that the richest 22 men in the world owned more wealth than all the women in Africa. Five of the 15 UK prime ministers since 1945 were educated at one very expensive boys' school: Eton College. All 15 were white, only two were women. Of the last six American presidents, two were father and son, and another was the son of a multi-millionaire. And the decision-makers at those fossil fuel giants? In 2019, only 14 per cent of oil and gas company board members were women, according to S&P Global.

With causation comes denial. We've seen the evidence on fossil fuel companies, carefully curating scientific doubt. Sociologists Aaron McCright and Riley Dunlap used ten US Gallup surveys, from 2001 to 2010, to confirm what might seem anecdotally obvious: white, conservative men are much more likely to deny climate change.

Privileges intersect. Income and wealth are inseparable on the one hand from race, gender, class or caste, sexual orientation, ability or disability. They are inextricable, on the other, from political power and climate complicity. Yet often, in mainstream talk of climate change, we blur distinctions, speak of 'our generation' or 'all humans' as responsible for environmental devastation. Take the 'Anthropocene':

the widely used term for the current geological epoch, in which human activity has dominated the planet. It's geologically appropriate, it's attention-grabbing, but it serves to perpetuate a fudge. '*We* are all to blame,' it suggests. '*We* must all act.' It makes no distinction between those who have caused climate change and those who are least responsible for it – and, as we will see, are most vulnerable.

We might do better, says ecofeminist theorist Giovanna di Chiro, to call it the '*WhiteMan* thropocene'.

CLIMATE CHANGE IS RACIAL INJUSTICE

Climate harm is about lack of privilege. It hurts those already disadvantaged worst of all. In sweeping terms, that means the global south. 'I can tell you,' writes Vanessa Nakate, 'a 2°C hotter world is a death sentence for countries like mine.' For millions of Africans, 1.2°C of warming is a living hell: a life sentence of depleted food, cyclones, floods, droughts.

In 2014, the IPCC warned of escalating heat-related deaths in Asia, widespread damage to infrastructure and livelihoods, floods, droughts and malnutrition. In Africa, the forecast is for food insecurity, drought and pest damage, disease and floods. In 2018, the IPCC zeroed in on that crucial half a degree of warming: 1.5°C to 2°C. Above 1.5°C, extreme heatwaves will be widespread in the tropics. Small island states and least-developed countries will suffer the most.

Children are particularly at risk. They're more susceptible to malnutrition, asthma from polluted air, vector-borne diseases like malaria and diarrhoeal illnesses from dirty water. UNICEF warns of devastating impacts along South Asian coasts, the Mekong Delta, the Nile river basin, on small island states and the Latin American Pacific coast. In other words, it is mostly children of colour who starve, fall ill, are ripped from their homes by hurricanes, lose their families, and face violence, exploitation and trafficking.

Those in the firing line are also those with least sway on the international stage.

'It was painful to witness a group containing the leaders of a number of smaller nations literally shut out from the negotiations and standing in the December snow while the major world powers argued late into the night.' Those are the words of Mary Robinson, former Irish president, describing the 2009 Copenhagen climate summit. And the outcome? The

1.5°C target, which might have protected small island states, was quietly dropped.

Within countries, the disparity is just as stark. 'All air is not equal,' sociologists Phoebe Godfrey and Denise Torres tell us. And they're right.

Asthma rates are higher in the United States for African American and Indigenous children. Research by scholars including Jeremy Hoffman, chief scientist at the Virginian Science Museum, shows that non-white neighbourhoods in the US are up to 7°C hotter than others in the same cities. An industrial corridor between New Orleans and Baton Rouge is called 'Cancer Alley', for a reason. The average cancer risk from airborne toxins in the US is 30 cases for every million people, on 2014 Environmental Protection Agency figures. In Reserve, a working-class town at the epicentre of Cancer Alley, it's 500 in a million. The communities inhabiting this deadly strip are, mostly, Black.

Aboriginal people were devastated by the 2019/20 Australian bushfires but have been marginalized in most previous public responses. That's according to Australian National University researchers, led by Euahlayi scholar Bhiamie Williamson. In the US, pipelines carry oil to exacerbate the climate crisis, across fragile Indigenous lands. The Standing Rock Sioux Tribe believe the Dakota Access Pipeline threatens its very survival. Line 3, earmarked for expansion as I write this, would bring its toxic cargo through the North Minnesotan lands and waterways, where Indigenous communities hunt, fish, and gather rice.

Hurricane Katrina was a disaster, but it was not natural, and it was starkly racialized. The images are unforgettable. Black victims, left to die; Black Americans attempting to flee, barred from crossing to a predominantly white suburb. Six days after the hurricane, with the city still in chaos, New Orleans police shot four African Americans on the Danziger Bridge. Two were wounded, two killed. One had severe mental disabilities. The police said they were defending themselves, but were subsequently accused of a cover-up. Ultimately, five officers were jailed.

In 2007, Harvard sociologist Patrick Sharkey trawled the records of those missing and killed. Black Americans made up only 51 per cent of the elderly population of New Orleans, but 58 per cent of elderly victims were Black. Among the non-elderly, the figures were respectively 70 per cent and 82 per cent. Of those still missing when Sharkey wrote his article,

84 per cent were Black, compared to 68 per cent of the city's population.

CLIMATE CHANGE IS CULTURAL VIOLATION

'If the glacier were to disappear, I wouldn't lose my faith . . . but I would be heartbroken. A part of me would disappear.' That's what Aybar Quispe, one of the Indigenous dancers known as the 'guardians of the glacier', told *National Geographic* in 2021. The glacier is Colque Punku, in Peru. It's sacred to several Andean Indigenous tribes, integral to the Qoyllur Riti'i ('Snow Star') festival. And it is melting.

The last section described undeniable rights violations: elderly people drowning, children choking on polluted air. But something more intangible is being taken as well, and taken, by and large, from people of colour. A way of life.

Martha Nussbaum thinks we need to 'live with concern for and in relation to animals, plants, and the world of nature', to lead a full life. Whether or not we all agree with this – let alone feel the need to connect directly with non-humans – many people do. They care deeply about the places they live in and the species with which they share them. And, as humans, we value being able to live by something that matters to us: a spiritual or cultural framework. Maybe it's not a 'basic need', like breathable air and food to eat. But, as Quispe knows, it's life-limitingly painful to lose.

In small island states, people stand to lose their country, their homes, *and* their whole way of life. Indigenous communities are being stripped of what Kyle Powys Whyte, University of Michigan professor and member of the Citizen Potawatomi Nation, calls 'collective continuance': specific and valuable knowledge, the ability to adapt as a cultural community, so that members can flourish. Pipelines threaten culture, as well as livelihood. For the Inuit, in Nunavut in Canada, climate change undercuts location-specific survival skills and traditional hunting, fishing and gathering. It threatens historical sites. It makes it hard to pass on traditional skills for predicting the weather.

Like Quispe, the mountain-dwelling Kallawaya, in Bolivia, are sustained by their connection to the land. As the ice melts, everything is at stake. History, community, culture, unique knowledge. The Kallawaya are adapting, finding ways round, in this place that means so much to them. But as geographer Dylan Harris puts it: 'What happens when the people are forced to leave their mountains behind?'

THE LEGACY OF COLONIALISM

We have, then, a snapshot of injustice: those races, cultures and communities least to blame for climate change suffer most, and are kept out of decision-making. It's a true image. But a snapshot is only part of the picture. To fathom the depths of the injustice, we have to appreciate the power dynamics that got us here.

Let's look at the history. This is about the industrial revolution, the centuries from which elites in the global north have benefited, the political and economic power accumulating in the hands of a minority. But it's also about the flipside: colonialism, slavery. Unbelievable cruelty imposed by men and women on men and women. Without acknowledging this, we cannot understand why environmental harms fall along sharp racial lines.

States rich in fossil fuels do not necessarily benefit from industrialization. Thanks to the so-called 'Resource Curse', the reverse is often true. Ordinary citizens die, while shareholders in so-called 'more developed countries' – and a minority in resource-filled states – grow rich on oil.

Consider Ogoniland, in the Niger Delta: a case study of how not to do climate justice. In 1958, the Anglo-Dutch giant Shell began drilling for oil. For decades, spillages stripped communities of food, health and clean water. More than 60 years later – long after Shell itself withdrew from Ogoniland – the wait for reparation continues. Generation after generation has grown up breathing hydrocarbons and fighting for their human rights. In 1995, Ogoni leader Ken Saro-Wiwa, demanding justice, was executed on what were widely regarded as trumped-up charges. In 2021, the UN Environment Programme produced a plan to redress the devastation. And yet the title of a 2020 NGO report says it all: 'No Clean-Up, No Justice'.*

Tuvalu and Kiribati were scarred by slavery and colonialism, before they faced climate change. In the late 1800s, Peruvian slave ships devastated their populations; for most of the 1900s they endured British rule. Control over their own nations, secured for less than 50 years, is now vanishing under salt water. In North America, Indigenous communities fight oil companies just as they have protested, for centuries, against the annexing of their territories. In Bolivia, the Kallawaya were poor even before climate

*There is, at least, an encouraging postscript: in 2021, villagers from the Ogale and Bille communities succeeded in a five-year fight to bring their claims against Shell to the UK courts.

change took their remaining land. Wealth accumulates in cities, held by the descendants of colonizers.

'All air is not equal.' It's not equal for a reason. Those soaring temperatures in mostly Black communities? Less surprising when you know that decades of planning policy kept the green spaces for whiter, richer neighbourhoods, or that federal building programmes put major roads and construction projects in poor areas. Climate injustice feeds on institutionalized environmental racism.

The environmental justice movement sprang from within the civil rights movement, provoked by the radically disparate impact of pollution and hazardous industries. In 1991, the multinational People of Color Environmental Leadership Summit declared: 'Environmental justice demands that public policy be based on mutual respect and justice for all peoples, free from any form of discrimination and bias.' Public policy has yet to catch up.

Consider Louisiana, again: a state of sweeping plantations, of centuries of slavery. Now injustice comes in a new package: institutionalized exposure of African American communities to preventable disease, or floods. The story of Cancer Alley goes like this. Petrochemical factories are set up in poorer areas. Because of the long arm of slavery, of Jim Crow laws enforcing racial segregation, of prejudice and discrimination, those areas are mostly Black. Factory fumes kill people, slowly and painfully. The community protests, but because it's poor, and those vocal activists are often Black women, it's hard to get traction. The pattern is repeated, across the Southern states and for Black communities in the North.

After Katrina, some commentors said it was simply 'chance' that many victims were Black, given that New Orleans is a predominantly Black city. Patrick Sharkey, who examined the records, disagreed. Neighbourhoods with high death counts were, on average, those with more Black residents. The sea doesn't choose its victims, but existing patterns of oppression and exploitation put some people in the way of the rising tide.

There is a legacy of cultural imperialism, too: of ignoring, devaluing and destroying traditional forms of knowledge. With that comes the ultimate irony. Climate warnings, born of such knowledge, have long been ignored. For decades, Canadian Inuit have observed the ice growing softer, less stable, forming later in the year. They've said so. But politicians wouldn't listen, until Western science backed them up. Even then, they didn't act.

'WOMEN AS VICTIMS, WOMEN AS LEADERS'

Following the 2004 Indian Ocean tsunami, Oxfam surveyed villagers in Indonesia. In four villages in the Aceh Besah district, close to three times as many survivors were male as female; in the four villages in the North Aceh District, 284 of the 366 who died were women or girls. In one village, four females died for every male.

For Ugandan women, climate change means malnutrition, sickness – and domestic violence. Traditionally, men grow crops for sale, women tend plants and animals to feed the family. As climate change cuts yields, tensions grow over what to prioritize: women come up against not only their changing environment, but their own menfolk. They also forgo adequate nutrition themselves, to keep men and children fed. In Mali, droughts destroy crops and men leave in search of jobs. Women are forced to juggle their own work and men's, without the superior status and rights that being male confers. Around the globe, the lives and responsibilities of Indigenous women revolve around flora and fauna, land and water, putting them especially at the mercy of rising temperatures.

These were just some of the findings submitted in 2019 to the United Nations Framework Convention on Climate Change (UNFCCC). The 'overwhelming message'? Climate change is not gender neutral. Women are hit worse than men.

Meanwhile, women are particularly adept at responding to climate change. They lead from the front, at grassroots level. The Ugandan farmer Constance Okollet has gone from petitioning her local council for seeds, to setting up a credit union and speaking at UN climate summits. Vu Thi Hien, a former academic, has negotiated a hierarchical bureaucratic system and worked with Indigenous communities to keep thousands of hectares of Vietnamese forest safe from loggers.

These are just two women. Women I can name because Mary Robinson has used *her* privilege to tell their stories. There are countless others out there, driving climate action, shaping adaptation, often going unrecorded.

In 1991, Bella Abzug and Mim Kelber, respectively a former US congresswoman and a feminist activist, set up the Women's Environment and Development Organization. In the 1980s and 1990s, the environmental justice movement was led mostly by women of colour. Today's youth climate activist leaders are women in their teens and twenties, from Sweden's Greta Thunberg to Alexandria Villasenor and Jamie Margolin in the United States; from India's Licypriya Kangujam and Brazil's Artemisa Xakriabá, to the UK's Holly Gillibrand and Germany's Luisa Neubauer; from Canada's Autumn Peltier to Uganda's Vanessa Nakate and Leah Namugerwa.

Women get things done. According to research by economists Astghik Mavisakalyan and Yashar Tarverdi, the more women are represented in national parliaments, the more stringent climate policy is, and the lower carbon dioxide emissions.

CLIMATE CHANGE IN A MAN'S WORLD

Given all this, it's tempting to grasp at the trope that headed the previous section: 'Women as victims, women as leaders'. It's snappy, almost reassuring. 'We've got this,' it suggests, sighing but competent, like overburdened women the world over. But this is dangerous. It's insidious. It lets the pressure fall on individual women: to change their lives, make more sacrifices, be the 'environmentally virtuous' ones. The onus should be on governments, on corporations. It should be on everyone living comfortably, to reform the systems on which we all depend.

The trope is also misleading. There is only so much that women can do with the cards stacked against them. Yes, grassroots leadership is essential. Yes, women's voices are being built slowly into institutions. For example, the Women and Gender Constituency is one of nine UNFCCC stakeholder groups. But even resilient pioneers can be stymied by institutional barriers to resources or information. Grassroots leaders don't generally get a say in national or global decision-making, and women remain woefully under-represented at the top levels. In 2020, the UNFCCC's decision-making and technical bodies included only 33 per cent of women on average, according to its own figures.

Moreover, this is not just about getting women into decision-making roles. It is about the culture in which decisions are made. According to some feminists, including the writer and activist Naomi Klein, the whole agenda is dominated by masculine thinking. This skews not only who is allowed to suffer from climate change, but which climate policies get airtime. Responsibility is pushed onto individuals to change their lives, rather than hold fossil fuel corporations accountable. Technical fixes, even dangerous ones, are given more airtime than serious institutional change.

Women suffer more than men from climate change, but not because they *are* women: because of what society has made that mean. And not *all* women suffer equally, just as not all men are equally advantaged. That's the other problem with the 'women as victims, women as leaders' trope.

Point one. This is not about biological differences. The UNFCCC report says that explicitly. From women in the global south searching further afield for fuel, food and water, to those facing escalating levels of rape and violence as oil and gas is extracted from Indigenous land, climate change harms women in particular ways. But it does this because of how women's roles have been collectively defined, and their interests subjugated.

Here's another true story of structural disadvantage. 'Lialeese',* an Ethiopian woman in her thirties, was attacked with an axe while she was guarding the forest from illegal loggers. Disfigured and in agony, she was denied medical treatment because her husband refused to pay for it. 'Lialeese could not speak due to her injury,' wrote Victoria Team and Eyob Hassen, the researchers who told her tale, 'but she could

*Not her real name.

also not be heard due to the lack of women's voices in Ethiopia.'

'Women in the South', says a 2011 UN Environment Programme report, 'are particularly vulnerable to the impacts of disasters due to skewed power relations and inequitable cultural and social norms.' If a woman has no control over the resources she needs to live, she cannot adapt. If girls can't learn to swim, or to climb, they are at the mercy of floods. Women, weighed down in water or fire by long, heavy garments, are literally killed by gendered dress codes. If they escape, it is to face further risks. Violence, intimidation, sexual harassment, rape, organized trafficking.

Point two. If it's no coincidence that women are harder hit by climate change than men, it is also no coincidence that the women we are talking about – Indigenous women, climate refugees, women fighting floods in the global south, Ethiopian forest guards – are not usually white.

CLIMATE CHANGE IS INTERSECTIONAL INJUSTICE

'When sorrows come,' writes Shakespeare in *Hamlet*, 'they come not single spies, but in battalions.' In our context, they do so for a reason.

The term intersectionality was coined by Kimberlé Crenshaw, a Black feminist, lawyer, philosopher and civil rights advocate. In thinking of the injustices perpetuated along race and gender lines, she says, it's easy to ignore the unique experience of women of colour. Their pain is often untouched by policies that benefit only white women, or men of colour. Intersectionality allows us to understand how someone's social and political identities combine to silence them, burden them with discrimination, or entrench their privilege.

Lialeese's human right to healthcare was set aside not simply because she was a woman or in the global south, but because she was a woman in Ethiopia: a society already globally disadvantaged, and itself structured to make women vulnerable. Indigenous women were *already* more likely to suffer violent crime, before fossil fuel employees made things even worse. Climate injustice is race injustice. It is gender injustice. But it is something else too, reducible to neither.

Other differences matter as well. According to geographer Aleksandra Kosanic and colleagues, people with disabilities are more vulnerable to disease or extreme weather. A 2019 *Grist* article describes young queer and trans people in Jamaica rejected by their families, driven by discrimination to makeshift camps outside the city. As climate change exacerbates extreme weather, they are at its mercy.

'My partner and I both needed our meds, clothes and a way to find permanent shelter after the storm,' Jeremiah Leblanc, a gay, Black survivor of Hurricane Katrina told the *Huffington Post*, 'but we knew to stay the hell away from the black churches offering help. We couldn't tell anyone we were sick and HIV-positive. And when we got to Houston, we saw the Salvation Army, but [we] knew to stay the hell away from that too.'

Without understanding these differences, policies backfire. Compare my life with that of a woman in Bangladesh, made homeless by climate change, living hand to mouth in a Dhaka slum, constantly afraid for her children. Yes, we're both women. We have a lot in common: we love our families; we need certain things for a decent life. But I have those things, and she doesn't. Policies that would help me, by redressing the ongoing inequalities of UK society, wouldn't scratch the surface of her needs. To treat the victims of a disaster like Katrina as a homogeneous group – even the 'Black victims' or the 'women victims' – is to overlook those who are most in need.

One of the *Grist* authors, 20-year-old Phillip Brown, was a Black, queer immigrant who fled violence in Kingston. They joined the youth climate justice movement in the US to speak up for those whose experience is too often ignored. If communities are represented only by straight people without disabilities; if the representation of women in climate decision-making is, in practice, the representation of *white* women, women from Europe and North America – or even of just a few women cherry-picked from southern governments or universities – then climate vulnerabilities will never be understood.

Six months after Nakate was cropped from that image which flashed around the world, another climate justice advocate of colour, Mary Annaïse Heglar, wrote: 'It's time to talk about my biggest fear about the climate crisis. It's not, "How will we treat each other?" It's "How will white people treat people who look like me?"'

Climate change violates human rights. We saw that in Chapter One. But it doesn't violate them at random. Climate injustice feeds on existing violence and discrimination, on a history of rights abuses. Until we understand that, there is no prospect of climate justice.

3

Beyond Humans

The lemuroid ring-tailed possum. A small symbol of a vast and frightening phenomenon: climate extinction.

Google it, and you'll see dark eyes in dense fur, a marsupial perhaps 30cm long with a tail as long again. You'd find them in the Australian Wet Tropics, high on the slopes of Mount Lewis or the Atherton Tablelands in Queensland. Sleeping in tree hollows, waking at night to jump tree to tree in the rainforest canopy, living their own lives.

Only you very likely *wouldn't* find any, because soon they will have no life to live. After the 2005 heatwave, the rare white variety on Mount Lewis vanished for several years. According to the *State of Wet Tropics* report, the entire southern population has dropped by nearly three-quarters since 2008. Climate change, heating the habitat of these tiny creatures, drives them higher and higher into their native hills. There is only so far they can go.

In 2016, another Australian, a tiny rodent known as the Bramble Cay melomys, acquired a dubious distinction. It became the first mammal to be wiped out by climate change. Sea turtles, Antarctic penguins and Hawaiian honeycreepers are all heading the same way. The polar bear, poster child of climate devastation, could disappear by 2100.

This is just the tip of the iceberg. Or would be, if we weren't melting them. As extremes become normality, prospects are bleak for many non-humans, species and ecosystems. That's an injustice too.

Let's be clear, before this chapter gets underway. It wouldn't *have* to be an injustice – this mass harm to non-humans – for us to know that climate change is morally atrocious. The last two chapters make that clear enough. Global warming is terribly, avoidably bad for human beings. Perhaps you are convinced that non-humans *don't* matter, that you can ignore the impact of climate change on them, that we have enough to worry about thinking about humans. Even if that's the case, I'd ask you not to skip this chapter. It turns out to be more complicated than that.

MASS EXTINCTION, MASS DEVASTATION

The short story: global biodiversity will be devastated, unless we go all out to stop climate change. The

World Wide Fund for Nature (WWF) has identified 35 'priority places' for conservation. If temperatures rise 2°C, only 56 per cent will stay suitable for the species inhabiting them. With a 4.5°C rise, that plummets to 8 per cent. According to the IPCC, a 2°C rise would wipe out more than half the liveable range for 18 per cent of insects, 16 per cent of plants, 8 per cent of vertebrates. Even at 1.5°C, these figures are non-trivial: 6 per cent, 8 per cent, 4 per cent respectively.

Take the Miombo Woodlands, an area spanning 2.4 million square miles of central and southern Africa. With 2°C of global warming, 45 per cent of mammals, 47 per cent of plants, 48 per cent of birds, 54 per cent of amphibians, and half of reptile species could disappear. For bird and mammal species, there's some hope of escape through moving to more hospitable territory. For plants, amphibians and reptiles, there is none. At 4.5°C, more than 80 per cent of all species are at risk, 90 per cent of amphibians. In the Amazon, a 2°C rise puts more than a third of species in danger. A 4.5°C rise and almost three-quarters of amphibian species could be decimated.

Behind these figures lie countless incidents of individual suffering. Species go extinct because individual animals cannot survive. They die, *every single one*. They starve. They drown. They die of heat exhaustion, or because they have nothing to drink. They burn to death. Nearly three billion animals were killed or displaced by Australia's summer of wildfires in 2019/20. According to the WWF, many who escaped the line of fire died of injuries, deprivation and stress, as more and more animals crowded into unburnt areas.

We've all seen the headline images. Polar bears, shabby and desperate, raiding northern cities. Koalas, black and red with burns. We wince, turn away. Most of the victims we never even see. But, on plausible philosophical models, that harm matters too. Let's talk about why.

'CAN THEY SUFFER?'

As humans, we owe it to each other not to do serious harm. We should go further: spare other people from suffering, protect their basic rights, if we can do it relatively easily. That was spelled out in Chapter One. It is, I hope, uncontroversial. But what is it about human beings, each and every one of us, that gives us this moral significance?

Think about it. Take your time. Whatever capacity you come up with, you'll have to do one of two things. You'll find that some non-humans have that capacity too, so we have moral duties to them, or you'll pick something that not all humans share, which would mean some humans aren't entitled to the protection of morality. Peter Singer calls this the Argument from Marginal Cases. Another philosopher, Jeff McMahan, refers to the 'Separation and Equality Problems', because we can't have it both ways: maintain that all humans are morally equal, and keep the clear line separating us from other animals.

Take rationality. Immanuel Kant theorized that we don't owe non-humans anything because they are not capable of rational thought. He proscribed animal cruelty, but only because torturing animals might be a stepping stone to ill-treating humans. But not all humans are rational. Babies aren't. Some people with severe intellectual disabilities aren't. Does it not matter if *they* are harmed? The idea is repellent. If anything, their greater vulnerability makes it worse to allow them to suffer, more contemptible not to protect them.

So perhaps it is exactly that feature which matters: suffering. Our ability to experience physical and mental pain makes us worthy of moral concern. We *can* all suffer. But we're not the only ones.

Non-human animals experience pain. They feel fear. They suffer in myriad routine ways at the hands of humans. Through animal testing, through blood sports, in meat and dairy industries. Because of climate change.

As the philosopher, social reformer and founder of utilitarianism Jeremy Bentham put it, 'The question is not, Can they reason?, nor Can they talk? But, Can they suffer? Why should the law refuse its protection to any sensitive being?'

In fact, pain is simply the bottom line. Other animals have more in common with us than that.

Danielle Celermajer is an Australian philosopher who runs a smallholding of rescue animals. In 2020, still reeling from the fires, she wrote a book, *Summertime*. Read it, and I defy you not to marvel at the capacities of our fellow creatures. Celermajer bears witness to this in better times, recalling the friendship, communication and complex personalities of donkeys, and in terrible times. She describes the deep grief of her pig, Jimmy, whose companion died horribly. He is traumatized, 'hypervigilant'. He searches for her. He sniffs about, smelling for her. He sinks into desperate stillness. He doesn't eat. He mourns.

Cognitive science bears this out. The lives of non-humans can go well or badly, and there is a sense in which they are aware of this. Take the findings of animal welfare scholar David Mellor. Many non-humans can experience anxiety, panic, frustration and anger, as well as fear and pain. They can feel bored or helpless. They can be depressed. They can also have positive experiences: comfort, pleasure, being energized, being engaged, being 'affectionately sociable'. They can be fulfilled by parenthood. They can feel nurtured and secure. They can be excited and joyful and gratified.

The implications are terrifying. Every one of those animals roasted alive by wildfire or tortured in factory farms has the same capacity to live and thrive, to form relationships and feel joy, as the two rescue pigs whom Celermajer loved, and makes her reader weep for.

If this is injustice, the scale is mind-blowing.

WHAT JUSTICE MEANS

But *is* it injustice? What follows from this commonality? If it is wrong to harm other people, it must also be wrong to hurt our fellow creatures, at least the sentient ones, whether individually or through our governments, corporations and global food market. Do we also have a duty to *protect* them, to make sure that each and every one can lead a decent life? For consistency, it would seem that we do.

Martha Nussbaum agrees. Sentient individual non-humans, she says, deserve capabilities-based justice. Pigs, sheep, dogs and dolphins, and others like them, have a claim on us: not only not to be treated cruelly, but to the 'prerequisites of a dignified existence'. Food and space, light and air. Freedom from squalor or fear. Freedom to interact with one another, to run, to climb, to play, to flourish *as* the animal they are.

Perhaps, like the hypothetical objector in Chapter One, you want to stop me here. You want to bang your hands on the table and tell me firmly that we're talking about *justice*. Justice, you want to say, isn't about being 'kind' to animals. It's about what we owe one another as reciprocating members of a society, playing by rules we would all have agreed on. My cat, you point out, is not a member of society in the same way that I am. Nor is a cow. They can't pay taxes. They can't vote.

Well, obviously they can't. But bear with me, because that doesn't undermine what this chapter is saying.

Take that objection as it stands, and we face the problem that Peter Singer and Jeff McMahan have posed. Plenty of humans can't do those things either. Babies can't vote, pay taxes, or even work. Nor can the severely intellectually disabled. They can't reciprocate if that means playing an active part in the cooperative institutions of the state. Are babies not entitled to the protection of justice? Are the most vulnerable adults to be set aside? The thought revolts.

What's more, non-human animals can reciprocate, in a sense. They *are* part of institutions. They live and die on farms, in cages, in science labs. They share our lives as companion and support animals. Remember what Chapter One asked, about humans in another state or generation. Are those who are coerced by our institutional systems to be denied their protection, simply because they have no say in running them?

A final point, again channelling Chapter One. We're not talking about equality, for men or pigs. We're talking about what it takes to lead a decent life. We're talking about living, sentient beings systematically being deprived of core opportunities. We're talking about injustice, yes, but injustice at a very basic level.

TIGERS AND GAZELLES

That said, it's not straightforward to expand the obligations of justice like this. The implications go very far indeed, possibly *too* far.

Non-human animals are harmed by institutions: by us. They are subjected to pain and indignity to provide food, cosmetics and medicine. They are burnt and starved and drowned by climate change. But they are also harmed by each other, *all the time*. Thriving as a tiger means eating gazelles, which is horrible for those individual gazelles.

Martha Nussbaum has a solution. She thinks some capabilities don't merit protection: capacities for cruelty, for inflicting pain. When it comes to humans, this makes intuitive sense. But if she's right, we could have a mandate for wholesale interference with the natural world. Nussbaum stops short of advocating 'policing nature' (protecting prey from predators) but only because, in practice, it might well do more harm than good to individual animals.

For many, including conservationists, this misses the point. Intuitively, there is something wrong in seeking to 'sanitize' nature, however uncomfortable we feel about the suffering inflicted within it. Or think about it this way. When we mourn the losses caused by climate change, we mourn not only individual animals. Species vanish; ecosystems are destroyed. Isn't this a tragedy – an injustice, even – over and above the individual suffering?

On 19 March 2018, a 45-year-old died in Kenya. His passing was recorded in newspapers around the globe. His name was Sudan, and he had been described as the most eligible bachelor in the world. Sudan died peacefully, euthanized, after a last rub on the ear from his caretaker. Countless mammals died worse deaths – horrible, unrecorded deaths – on the same day. If what I've just said is right, their deaths were morally bad too. But Sudan mattered in a different way. He was the last male northern white rhino, and his species died with him.

Systems matter for individual animals. They matter for us. I'll come back to that. But do they matter in themselves? Is it *wrong* to force a species to extinction, or decimate a rare ecosystem to build a golf course? After all, why should we focus on dignity, as Nussbaum does? Why not seek to protect everything with interests, or with integrity?

Think of a river system, as the political theorist David Schlosberg does. The species it maintains, how it functions as an integral whole, how that is disrupted when we put plastic or chemicals in or take water out. Think of England's chalk streams, vibrant with the blue flash of the kingfisher, the elegant flag iris. Home to mayflies and trout, salmon and otters and water voles. At least, that's how the streams should be. But water is taken from them, for people to drink and shower and send in hosepipe arcs over manicured lawns. Land is drained, human-built barriers block fish from passing along streams. Sewage and agricultural pollutants seep into the water. Heat spells, exacerbated by climate change, dry them further. Algae spread, fish die, the system is undermined.

Chalk streams, like us, can be healthy or otherwise. And now, according to the WWF, they are not healthy: by 2014, only 23 per cent in England were in what the Environment Agency classifies as 'good health', 46 per cent were in 'moderate health', and 30 per cent in 'poor health'.

Think again of those tigers and gazelles. Predation harms individuals, but it is how species maintain themselves and ecosystems thrive. Or go bigger again: think, ultimately, of the whole complex, interconnected ecosystem that is our Earth, and how climate change disrupts *that*. How whole species are wiped out.

JUSTICE AND LEAVING ALONE

If Schlosberg is right, ecosystems and species are entitled to have their flourishing respected. I suggest that we accept this. But what, in practice, does it imply? How are these different interests to be reconciled?

Let's start with this: Humans don't need to kill and maim each other in order to flourish; non-humans do. What's more, we are moral agents, to whom moral rules apply. This means that non-humans, if they have claims of justice at all, are entitled to protection *from us*. We can also draw a distinction, as several philosophers have, between domesticated animals, already under our control, and individuals and species 'in the wild'.

For domesticated animals, Martha Nussbaum's protections look like a bare minimum: the basics for a dignified existence. Food to eat, space to roam, air to breath. Not enduring dark or squalor or fear. Being able to form bonds, act as the kind of animal they are. We also become responsible, in a way, for the actions of domesticated animals. We put bells on pet cats, keep them in at night; we muzzle dogs liable to attack others; if we must have zoos, we make sure the predator cages are *very* well fenced.

To wild animals, ecosystems and species, we owe something different. We should respect the integrity of complex wholes. Put simply, we must leave them alone.

Yes, conflict remains. Reality is harsh. The individual gazelle suffers, killed by a tiger. Her *individual* interests are undermined. There is no way to reconcile this. But it's not our job to reconcile it, because individuals also need the systems that predation maintains. Disrupt one part of that coherent whole (remove or disarm tigers) and you undermine another. (Too many gazelles, not enough food.) To live as a gazelle is to live within this vibrant, beautiful, sometimes horrible, whole.

So far, so promising. Except that it is centuries too late for non-interference. Quite apart from all the other ways we have wrought devastation on systems and driven species to extinction, how do you draw a line between 'natural' and 'unnatural' when we're heating the whole planet? Even many of the places valued as wild and beautiful are marked by past human actions. Ecologically, the Lake District and Scottish Highlands are barren echoes of what they should be, because of sheep farming.

What, then, is 'wild'? What is just?

Here's a suggestion. Doing justice to wild animals, species and systems means preventing and correcting harms done by humans, ending and alleviating the suffering inflicted by structures that we have created and of which we are a part. This isn't about saying that no one should help koalas decimated by wildfires, or support Amazon birds and mammals to rebuild their populations elsewhere. Such interventions can be conducted in the spirit of justice: as an attempt to rectify previous human interference.

In practice, it's hugely complex. Trade-offs are real, choices are difficult. We'll talk about them more later. But that's the basic idea.

What's more, pervasive as human influence is, we should not overstate it, or treat it as justification to interfere everywhere. There's still 'naturalness', says the philosopher Ned Hettinger. Species have lost elements of natural behaviour, but other aspects remain. Polar bears, driven to hunt in cities, still seek sea and seals when they can. There is a huge difference between African elephants roaming on even depleted savanna, and in a safari park. There is still something worth preserving.

'US AND THEM' . . .

Let's return to humans.

We cannot separate the fates of humans and other species. This is obvious. It is the failure of much of the modern world to understand this – the tendency to put *man*kind on a pedestal – that has landed us where we are now. (I use the male pronoun deliberately. I could perhaps have said 'white mankind'. But I'll come back to that.)

We need the natural world. That's true whether or not we depend psychologically on our interactions with it: on the spiritual sustenance of religious sites and totem species, the sheer pleasure of running through woods. Animals can be members of our communities. We saw that already. But the non-human world, as a whole, enables us to live. Trees give us oxygen; plants and animals provide our food. We depend, utterly, on interconnected ecosystem services, on biodiversity outside (and, for that matter, *inside*) ourselves.

'The Amazon's ecosystems,' says the WWF report, 'host around ten per cent of all known species and play a crucial role in regulating the global climate.' They're called the world's lungs for a reason. Climate change, worsened by the devastation of this beautiful system, in turn attacks what is left, and makes everyone more vulnerable again. This wholesale devastation of ecosystems, this biodiversity massacre, is *terrible for us too*.

That's why some philosophers think there's a specific human right to a safe or adequate environment. Others say it's a central interest or capability. The political theorist Breena Holland describes it as a 'meta capability', on which all the others depend, now and into the future. Her reasoning is simple: the natural world is so fundamental to our well-being that it breathes life into each aspect of a full human life.

In 2021, parent climate activist groups ran a global social media campaign, #ourothermother, using pictures and poems. It was both poignant and arresting. A baby in a sling, head nestled against the blue-green Earth. A woman's face, sad and serene, clutching the Earth, the picture darkening at the edges. A child, silhouetted, on a sinking world. The campaign was deliberately emotional, stressing the intergenerational. But it made precisely Holland's point. If we care about other people, if we care about our own children, if we care about *ourselves*, we have to care about the non-human world.

. . . BUT NOT ALL OF 'US'

All this is true. But just as the 'us' and 'them' dichotomy gets it wrong, so does this convenient, all-purpose 'us'. In recognizing the vulnerability of the non-human world and our dependence on it, we mustn't forget Chapter Two. It isn't all humans who cause climate change, and we aren't equally vulnerable.

There's a myth, says Mary Annaïse Heglar, that the Black community isn't interested in environmental issues, or protecting other animals. They mostly are. They're not so keen, Heglar says, on environmentalists. There's a reason for that: so-called nature-lovers, often older white men, have been known to sweep intersectional climate injustice under the carpet, present a 'this is all of our problem' narrative, then complain if Black people are more worried about police violence than polar bears.

Yes, Heglar says, climate change is an existential crisis. But, for people of colour, it's by no means the first.

Her article on nature-lovers has a brilliant subtitle: 'It's time to stop #AllLivesMattering the Climate Crisis'. In shifting the climate activism tagline to #All*Life*Matters, we must not give those in power more leeway to marginalize the interests of people of colour.

Here's another grim term: colonial conservation or 'fortress conservation', expelling Indigenous peoples from ancestral lands, in the name of protecting nature.

In a video for campaign group Survival International, a Baka man from the Congolese rainforest describes an attack by wildlife rangers. 'They started beating everyone.' An old man died, he says. A child 'about this small' died. He holds out a hand, to illustrate. It's on a level with his head as he sits down. The child was the height of my six-year-old. In 2020, the *Guardian* got hold of a leaked draft of a UN Development Agency report, finding 'credible' evidence of years of violence and abuse. If Survival International is right, these 'eco guards' are partly funded by the WWF.

The Ogiek people have a right to live in the Mau forest of Kenya, a position upheld in 2017 by the African Court on Human and Peoples' Rights. But the Kenyan government, which owns the land, has evicted hundreds of Ogiek families. Or so the *Independent* reported in 2020. The alleged aim? Conservation.

So is there another way? I hope so.

MULTISPECIES INJUSTICE

Here's the thing. Gender injustice, race injustice and inter-species injustice are part and parcel of the *same phenomenon.* Ecofeminist philosophers have been pointing this out for years. Nature is exploited by the same processes that oppress women and people of colour. 'Colonialism of the Earth', as the philosopher and ecofeminist Val Plumwood put it.

For generations, Plumwood says, those in the global north were spun the same old story: white men as rational, women and other races as 'irrational', 'closer to nature'; further (so the implication went) from any claim to superiority.* White men saw themselves as entitled to take control. If powerful voices like theirs resist the idea of being *part* of the natural world, that's because they have too much at stake. Or think they do.

And the way forward? Of course, it means recognizing the equality of all humans. But not by simply shifting the divide, drawing a sharper line between humans and non-humans, rational versus irrational. Women and people of colour *are* rational, but we can point that out without keeping rationality on its pedestal. Care and connectedness are valuable too. Nature matters. Progress means being humbler and more respectful, with each other *and* with other species.

*Plumwood herself got a bit *too* close to nature. She was mauled by a crocodile and almost drowned in the process. But being an incredible woman, she turned even that into a philosophical learning experience.

The activist Leah Thomas supports what she calls 'intersectional environmentalism', inspired by Kimberlé Crenshaw: advocating for vulnerable communities *and* non-humans, recognizing interconnected injustices. Thinking along similar lines, a group of thinkers founded by Schlosberg and Celermajer has coined the term 'multispecies justice'. This is justice for all species, races, genders, ages and abilities. But it's also intersectional in a broader sense. It recognizes that human individuals and communities, non-human animals and species, all live – and *can only live* – within a functioning environment.

Think like this, and justice for non-humans is not so much an 'extra' question as an inseparable part of the bigger picture. Two centuries ago, the same Clearances that paved the way for sheep to destroy Scottish ecosystems drove crofters from homes and country, to seek a place in lands taken from *their* own Indigenous people. Today, the rainforest is destroyed to produce cheap beef. What that does to human communities, present and future, is unjust; what it does to other species is also unjust. The one cannot be understood, or addressed, without the other.

4

What Climate Justice Looks Like

Whatever justice may positively require, it does not permit that poor nations be told to sell *their* blankets in order that the rich nations may keep *their* jewellery.

Henry Shue, 1992

On 11 June 2001, President George W. Bush refused to ratify the Kyoto Protocol, an international treaty on climate change. Standing against lavish greenery and an American flag, he announced that the United States wouldn't be part of a policy that wasn't based on 'global cooperation'. He was crying unfairness: China and India, major (but poorer) polluters, were exempt from Kyoto, and he objected to the US having to take on burdens if they didn't.

His decision was politically expedient; it was also morally wrong.

We have seen what climate injustice looks like: human rights are undermined, the prospects of future generations are erased. People of colour, women, Indigenous communities and other species pay a terrible price for a way of life from which a rich elite reap the benefits.

But what would climate *justice* look like?

WHAT JUSTICE DOESN'T LOOK LIKE

Let's begin with some ideas that might masquerade as climate justice, but aren't.

Justice doesn't mean allowing those who have historically produced the most greenhouse gas emissions to carry on emitting the most. (Yes, this is an actual proposal. It's called 'grandfathering': a term borrowed from policies denying the vote to African Americans at the turn of the twentieth century.) Allowances should sometimes be made for those hooked on a particular resource: if, perhaps, a person's actions aren't actively harming others, or he became reliant against his will. But this doesn't apply to high-emitting states and corporations who have raked in profits from destroying lives, and who have known for decades what they are doing.

As for affluent individuals, of course anyone living a luxury lifestyle on the back of fossil fuels has an interest in not giving that up (although perhaps not as big an interest as they might think). But that doesn't compete with others' basic rights. It's inconvenient to give up two holidays a year in the sun, but it's a lot more inconvenient to drown, or burn, or starve to death.

Another proposal is this. Everyone gets an equal share of any greenhouse gases that can still be emitted while avoiding 'dangerous' climate change. (In other words, a share of not very much. At the start of 2018, the global 'carbon budget' was less than 580 billion tonnes, and that, according to the IPCC, still only gives us an even chance of capping global warming at 1.5°C.) Alternatively, everyone gets an equal share of what's known as 'ecological space'. This includes your total use of natural resources: those you consume directly, and those used or destroyed in the process.

This is less unjust than grandfathering. It's even a step in the right direction, since it would mean huge emissions cuts by the rich. But it's still flawed. It doesn't allow for the vast historical differences in emissions patterns, the harm done and benefits reaped through climate change. Proposing this as 'fair' now is like a group of kids stealing half the sweets in the class jar, using the extra energy to win every race on sports day, then insisting that the remainder be shared equally between them all.

MITIGATION, ADAPTATION, COMPENSATION

Climate change does terrible damage: to today's children and *their* descendants; to millions of already vulnerable people, the world over. Climate justice requires preventing that harm, making up for it as far as possible. That takes mitigation, adaptation and compensation.

Mitigation means curbing global temperature rises. It means cutting greenhouse gases. A wholesale shift to plant-based eating, or away from flying, is part of mitigation. Transforming domestic heating from gas to heat pumps, powered by renewables, is part of mitigation. Leaving fossil fuels in the ground is essential. Mitigation can also mean *removing* greenhouse gases from the atmosphere. Carbon capture and storage is mitigation. Reforestation is mitigation.

But mitigation, now, won't be enough to protect human lives (never mind non-human ones). The IPCC is clear on this. Any global warming can damage human health. From 1.5°C, temperature extremes increase, droughts worsen, floods and tropical cyclones are more likely. By 2019, global temperatures were already up by just over 1°C, exacerbating poverty, increasing disadvantage.

We need to adapt. Adaptation means adjusting systems and institutions to protect lives and livelihoods from climate change. It includes insurance and education, adjusting infrastructure, finding ways to produce food using salt water, developing early-warning technologies for extreme weather. Adaptation is floodwalls in New York, water management in Durban, a floating farm in Rotterdam, crop management in India, cyclone shelters in Bangladesh. It's a heatwave response plan in Sydney, flood defences in Edinburgh. *Just* adaptation is – or should be – about prioritizing vulnerable communities.

Adaptation, too, can only go so far. Take the IPCC again. With 2°C of warming, there is a 'very high' risk of heat deaths in Asia – and 'medium' risks across categories and regions – even *with* adaptation. With 4°C, it's a lot worse. Even if societies try to adapt, people will still get sick. They'll suffer from extreme heat, damage to infrastructure, water and food shortages, and flooding.

Then there's compensation, for loss and damage, and other harms. A last resort, but a morally necessary one. Where the harm is done, compensation tries to make up for that in some other way, often financial. If mitigation is not pushing someone off a bridge and adaptation is building a safety rail, so he won't fall if you try, compensation is buying him a wheelchair when you've broken his back. For small island states, even 1.5°C of warming will push the limits of adaptation. If those citizens lose their homelands, their whole way of life, the loss can never be erased.

AGAINST DISCOUNTING

Some economists reject mitigation. They don't deny the grim toll of climate change on human lives, but they put it into the cost-benefit grinder and claim the economic risks of mitigation outweigh it. It is better, they think, not to take on the costs of preventing climate change, but to leave future generations to adapt to a hotter world. They'll be richer than us, the argument goes, so they can afford it.

There are various problems with this argument, but let's cut to the moral core. It's implicitly utilitarian, which means it's driven by *overall* welfare. It assumes that doing good to some people (including economic good) can outweigh doing harm (even very bad harm) to others. That's incompatible with the moral baseline we started with. Seriously harming other people is just plain wrong. What matters is decent lives, not growth at all costs.

The argument also relies on discounting: valuing future gains or losses less than present ones. And, as Simon Caney points out, this isn't just about standard economic discounting. It's about what's called the 'pure time preference': the idea that the further in the future people live, the less moral value they have. In other words, to make anti-mitigation sums add up, it must matter less if human rights are violated in 100 years than if they are tomorrow, in 500 years rather than in 100. Go far enough in the future, and they will barely matter at all. This seems very wrong. Remember Henry Shue's example, planting landmines for fun. Whether they'll go off in 100 or 1,000 years, it's still abhorrent.

Caney thinks this 'moral' discount rate should be zero, but sums come out in favour of mitigation even if it's positive, but very low.

We can start, then, with this. Climate justice requires mitigation, adaptation and compensation. But huge issues remain. For one thing, it's crucial to separate the following questions: who mitigates climate change, who adapts to it, who is compensated; and who pays for all this? For another, it matters, crucially, *how* we do it.

THE END DOESN'T JUSTIFY ALL THE MEANS

Mitigation is always a step towards reducing climate change, but not always towards climate justice. There are some obvious co-benefits. In 2014, the IPCC picked out opportunities to improve health *while* cutting greenhouse gas emissions: energy efficiency and cleaner energy; less meat and dairy, more pulses; better public transport, more biking and walking, fewer cars. Other policies, however, can themselves be fundamentally unjust.

Two such? Geoengineering and population control.

Put at its simplest, environmental damage is caused by a combination of how many people there are, how well-off people are, and how good they are at living well-off lives without scuppering the world they live in. In other words: Impact = Population × Affluence × Technology. This is the so-called IPAT equation, formulated by biologist Paul Ehrlich and colleagues in 1972. In theory, climate change could be mitigated by working on any of these. Even by geoengineering, a particularly radical version of 'T'. But there are moral limits. And the IPAT equation, so apparently obvious at face value, shouldn't necessarily be taken *at* face value.

'MENDING THE SKY'

Geoengineering, in the words of the Oxford Martin School, is 'deliberate large-scale intervention in the Earth's natural systems to counteract climate change'. It includes strategies to remove carbon dioxide, or to prevent accumulated greenhouse gases from heating the Earth. Tree planting, on a massive scale, is geoengineering. But so is solar radiation management, which tries to reflect some of the sun's energy back into space.

There's a huge difference between these extremes. One works with proven methods, seeking to treat the cause of climate change, not tinker with the symptoms. The other *sounds* great. Scientists inject sulphates into the stratosphere, or spray saltwater onto clouds. They make clouds more reflective; they put mirrors in space. And, boom, they've 'solved' climate change. But if it sounds too good to be true, that's probably because it is.

The philosopher Stephen Gardiner has warned repeatedly against solar radiation management, particularly stratospheric sulphate injections. It's fundamentally risky, unproven technology, with the planet as guinea pig. It commits future generations to carry on what we start, or risk dramatic, disastrous warming. In fact, it's so dangerous that pretty much only one argument for it could be made to work: that it's the 'last resort'.

But this, Gardiner points out, is disingenuous. Collectively, there are other ways of mitigating climate change: cutting greenhouse gas emissions, removing carbon dioxide. Perhaps injecting sulphates into the stratosphere is a lesser evil than unmitigated global warming, but that argument fails when the only thing standing between the global rich and non-evil alternatives is their own refusal to adopt them. Remember that hypothetical man on a bridge? Suppose you think it would be fun to push him off, so you rig up a net that might (but might not) catch him. You push him, then justify yourself by saying it was better than doing it without the net. Convincing? Hardly.

CLIMATE JUSTICE IS REPRODUCTIVE JUSTICE

Now think about 'P', population.

In 2012, the Oxfam economist Kate Raworth identified what she called the 'safe, just operating space' for humanity. This is a doughnut-shaped space, bounded on one side by planetary limits like climate change, and on the other by what's needed for adequate human flourishing. In 2018, the ecological economist Daniel O'Neill and colleagues warned that if the world's population increases as predicted by the United Nations (to nearly 11 billion, around 2100), the doughnut could become a 'vanishingly thin ring'. If they're right, our grandchildren or great-grandchildren's generation could face a tragic choice: one where all the options are morally terrible. A choice between their own central interests and those of the next generation.

This is deeply worrying. But we must not rush to dangerous conclusions. This generation's rich and powerful haven't used their opportunity to do basic global justice. The possibility of losing that altogether is no justification for doing more harm now.

Whatever this debate is about, it should *not* be about population as a scapegoat for climate change, biodiversity loss, the escalating violence that comes with depleted resources. But it too often has been. And let's be honest. That means women of colour as scapegoats. It means poorer states with higher birth rates, shouldering the blame. The 'population problem' must not be used as an excuse for coercive anti-natalist policies, closed borders, and the vilification of migrants.

The discussion should, instead, be about reproductive rights.

Point one. Population policies can violate basic rights.

In 2006, Reyhan, a dancer from the Xinjiang region of China, had an abortion. It was the second one demanded of her by China's one-child policy. That's bad enough, but doctors then sewed up part of her uterus, *without telling her*. She didn't find out, she told the *New Statesman*, until 2011. She had fled to Belgium and was trying to conceive a second child. For six years, she'd had shooting pains every time she sneezed, and hadn't known why.

One tragedy, among millions. (Literally, millions. In 2011, the official Chinese newspaper *The People's Daily* said its policy had 'prevented 400 million births'.) Nor did the violations end with the partial relaxation of the rules in 2013. According to a 2019 report for the US State Department, economic penalties and mandatory pregnancy examinations are still part of Chinese policy. Forced abortions and sterilizations may be less frequent, but they happen. And China isn't the only example. Mass sterilization has occurred in India; American women and children of colour were sterilized without consent as recently as the 1970s.

Other policies, designed to incentivize women to have fewer children, can be effectively coercive (as can policies to persuade them to have more). It's one thing to provide accurate information on how individuals can most effectively reduce carbon emissions, or challenge norms that make it hard for them to do so. It's quite another to fine vulnerable parents for having the large families they need to look after them in old age. That's not justice. It's completely the reverse.

Point two. The IPAT equation, widely criticized, is no straightforward case for population control. 'P' is just one of three variables, and not the most destructive. The appeal to average or overall affluence ignores the radically unequal distribution of wealth. According to research in 2020 by Oxfam and the Stockholm Environment Institute, even if the poorest 90 per cent of people in the world dropped their greenhouse gas emissions to zero tomorrow, the carbon budget would be used up only a few years later than it would otherwise have been. The richest 10 per cent would get through it all on their own.

It's terrifying to envisage a future where there literally isn't enough to go round. But we're not there now. The more stringently governments clamp down on fossil fuel corporations, the quicker the shift to renewables and away from fast-driving, frequent-flying, meat-eating ways of living, the less chance we will arrive at that grim point.

Point three. The IPAT variables aren't independent. Population doesn't grow because people 'just happen' to want lots of kids. Structural factors influence birth rates: that legacy of rights violations laid out in Chapter Two, extreme inequalities, maintained by institutions. And population growth can be slowed, helping to avoid a 'tragic choice' future, *without* threatening or forcing anyone to have fewer kids.

If infant mortality falls, if there are social provisions for old age, birth rates fall. If girls and women are educated and empowered, birth rates fall. If individuals are informed about family planning options and genuinely free to access them, birth rates fall. The IPCC knows this. 'Access to reproductive health services (including modern family planning)' is one of its 'win-win' opportunities. Such changes might be called 'population policy'. But they needn't be, since they are required by basic justice.

And, very probably, they shouldn't be so-called.

Make a policy explicitly demographic, the feminist Rosalind Petchesky warns, and it will be all about the target. Policies that cut birth rates as fast as possible won't optimally protect women's health. Hormonal contraceptives may prevent more unwanted pregnancies than condoms, but they don't protect against sexually transmitted diseases. What's more, healthcare alone isn't enough for reproductive justice, because it can't guarantee a genuinely free choice. Consider the complex social and economic reasons driving women to have (or not have) babies. What's the use of being legally free to have a child if you couldn't afford to raise it, or your work is so hazardous that it would be poisoned in the womb? A legal right to family planning doesn't help if you can't get to the nearest clinic, or if your husband would beat you if you did.

The lesson from this foray into the population debate? Climate justice means urgently prioritizing reproductive justice for its own sake – and welcoming the demographic side effects.

WHO SHOULD PAY?

Mitigation (with caveats), adaptation for all who need it, and compensation. All these are on the global climate justice tab. But who should pick it up?

In practice, any global deal must go via states. But states aren't the only relevant agents here. Consider corporations. The top 20 fossil fuel companies produced 493 billion tonnes of carbon dioxide and methane emissions from 1965 to 2018, on Climate Accountability Institute figures. Or consider individuals. States act on behalf of their citizens, protecting their rights and (sometimes) fulfilling their collective duties. Each state has richer and poorer inhabitants: those who fly weekly, and those who can barely afford the bus to work. It's just that there are a lot more of the former in so-called 'more developed' states.

With this in mind, there are three contenders for deciding who should pick up the climate tab.

Let's start simple: the polluter pays. Our most fundamental moral principle is a rule against doing harm to others, so *stop harming*, and clean up after yourself. That means mitigating, funding adaptation and compensating.

But this doesn't cover everything. What about any natural climate change? What about harm done by agents who are no longer around? Past generations, vanished corporations, even states that no longer exist.

Moreover, the no-harm principle comes with conditions. Don't do serious *foreseeable*, *avoidable* harm to others. How much climate devastation could have been predicted when the gases were first put into the atmosphere? Anthropogenic climate change became common knowledge in the 1990s. If we take that as the cut-off for 'predictability', plenty of harm is unaccounted for.

Some carbon emissions are unavoidable. A celebrity flying family and friends to a private island for a holiday could avoid this release of greenhouse gases into the atmosphere. A family living hand to mouth, cooking over a coal fire, cannot. A poverty-wracked state might have no option but to burn oil, because there's nothing else to live on. A multi-billion-dollar corporation makes the choice to dig for coal, rather than invest adequately in renewables.

In 1993, Shue distinguished between luxury emissions and subsistence emissions, or those we need for any kind of life at all. In mitigating climate change, it's the former that must be cut. It's not justice to expect anyone to starve. On the contrary, the global poor should be able to *increase* their standard of living.

Shue talked moral sense. He always does. Unfortunately, no one listened. Now we need to reach net zero emissions in less than thirty years, to keep global warming below 1.5°C. Even the very poor must live differently. But the core of Shue's argument remains. Everyone is entitled to a decent standard of living. The World Bank estimates that, post-COVID, more than 700 million people will live on less than $1.90 a day. Basic justice means raising that, *without* increasing emissions. Technology can help, but it needs to be paid for, and not by the poorest polluters. That's why it's essential for countries in the global north to develop clean technologies and transfer them to poorer states, or fund scientists in the global south to develop their own.

So the polluter-pays principle doesn't cover everything. There are also rich polluters who won't do what they should. Under the Trump administration, the US pulled out of the international Paris Agreement on climate change. Fossil fuel giants continue to extract oil and gas and coal. But these 'non-compliers' are so big a problem – their resolute indifference to even basic morality pulls us so far from any kind of ideal way forward – that they are a topic in themselves. I'll save them for Chapter Five.

PLUGGING THE GAPS

If this principle only takes us so far, why not turn to the second hard-to-reject moral rule that this book started with? If you can easily afford to prevent suffering, you should. We should protect other people's basic rights, individually or collectively.

The suffering isn't in question. Chapters One and Two made that clear. We look, then, to the rich to pick up any mitigation or adaptation costs not already covered. In academic terms, the ability-to-pay principle supplements the polluter-pays principle. 'The rich' includes everyone living a comfortable life, wherever they are. In country terms, it means the global north.

The costs of climate action are real. On 2014 IPCC estimates, keeping global warming below 2°C means reducing consumption growth this century by between 0.04 and 0.14 per cent, on average, per year. In 2019, the UK Committee on Climate Change put a price of 1 to 2 per cent of gross domestic product per year on reaching net zero by 2050. But these costs are not high in comparison to affluent standards of living: not high enough to justify leaving billions of people to suffer heatwaves, floods and disease.

The two principles so far have been straightforward applications of uncontroversial moral premises. But that's not the end of the debate. Doesn't it matter *why* the rich are rich? Even if those alive now aren't responsible for harm done by past generations, some of us benefit from it. Countries like the UK and the US became highly developed, and many of their citizens acquired income and wealth, as a result of industrialization. Those who live comfortable lives today – who are fed and educated and enjoy a multiplicity of opportunities – do so, for the most part, on the back of centuries of past generations burning fossil fuels.

Enter another proposal: the beneficiary pays. Those who gained the most from climate change (whether or not they caused it) should pay for mitigation, adaptation and compensation.

The core idea here isn't one of the two moral rules we started with. But it still gets at a widely shared intuition. Remember John Rawls' idea of justice as reciprocity. If we're part of institutions, if we benefit from cooperative schemes, we should do our share of making them work. In reaping the benefits of past harms and refusing to pay any of the costs, we're free riding. Or so says another philosopher, Axel Gosseries.

The libertarian Robert Nozick has an objection: 'I didn't choose this!' If I didn't ask for a benefit, I shouldn't be landed with the burdens that go with it. But there's a difference between trivial unchosen benefits and those fundamental to your quality of life. There's a difference between others taking on minor costs to provide those goods, and having their basic needs sacrificed, against their will. Crucially, we *know now* what price others must pay for our dependence on fossil fuels.

Here's an example. Two villages. In one, inhabitants enjoy swimming, fishing and feasting, right by a dam. As a result of the same dam, the other village, downstream, is waterless. You live in the first village. Your grandparents and their fellow villagers built the dam. Decades later, when they are all dead, you find out about the other village, how the inhabitants suffer, and why. Shouldn't this change how you act?

Or put it this way. Continuing to profit from past greenhouse gas emissions is like benefiting from stolen goods once you know they were stolen. It's a matter of climate debt (or, more broadly put, ecological debt: a concept long used by environmental activists). Maybe it's not fair to have to pay our grandparents' debts, if we're on the breadline. But if we're living comfortably on the back of huge burdens *they* pushed onto others, many of whom still suffer as a result? Then it seems reasonable.

A BALANCED PRINCIPLE?

Some philosophers think the beneficiary-pays principle stands alone, subsuming appeals to responsibility and ability and providing a single rule for divvying up the costs of mitigation, adaptation and compensation.

We needn't do that. After all, the other principles are grounded in the least controversial moral principles we're likely to find. But we can add nuance by incorporating the intuition against free riding into a hybrid approach. On Simon Caney's recommendation, polluters pay for the harm they've done, except if they are already very badly off. They pay for the harm they caused even if they didn't know it was harmful, so long as they benefited from it. The rich pay for the rest, especially those whose wealth came about unjustly.

This last includes not only those made rich by climate change – whose ancestors cut down forests and burnt fossil fuels – but also those aggrandized by other past injustices. At least, so Caney says, and I agree with him. In Chapter Two we saw how climate harms feed on historical abominations. It is only appropriate to demand greater redress from those who owe their privilege to violence and oppression. Climate justice cannot be achieved, or even understood, in isolation.

WHO GETS A SAY?

Climate justice also requires *procedural* justice. It's not just about getting to the 'right place' but about getting there in the right way.

As we've seen, it matters that people can flourish, or lead lives that go adequately well. But that means more than having enough to live on. Humans are autonomous creatures. We need to choose *how* to live, help determine the conditions within which we make those choices, and have our say on the same terms as everyone else. 'Control over One's Environment' is, Martha Nussbaum argues, a prerequisite for flourishing. That includes political control: effective participation in the political choices governing your life, free speech and association. Obviously, this means having a right to formal legal participation. It means having a vote. But it means more than that.

However morally plausible a model like Caney's is, translating it into practice is another challenge again. How should we calculate the benefits or shares of harm? How is any global scheme to be organized? Where are the greatest harms? What will it take to guard against them? All these decisions must be justly made.

In 2017, scholars compared climate adaptation plans from Australian councils with priorities identified by local community environmental groups. The disparity was stark. Councils focused narrowly on risk management, extreme weather events and legal liability. Local groups worried about day-to-day needs: food, water and energy. To know who is most vulnerable, and how, you need to ask those most concerned. Otherwise, harms cannot be prevented, or needs adequately met. *Just* adaptation means protecting the vulnerable from dramatic events: from storms, floods and fires. But it also means listening, learning what they need to get by in the 'new normal'.

This listening and learning could involve what's known as 'deliberative engagement': constructive exchange of ideas between citizens, scientific and social science experts, and policymakers. In 2014, academics ran a Citizen's Panel on adaptation planning in collaboration with City of Sydney representatives. Climate Assemblies have been held in Ireland and the UK. Such initiatives can be a big step towards climate justice, *if* – and it's a big *if* – policymakers act on the recommendations.

But, at the local and national level, they are still only a step. Climate change, remember, crosses community and state borders. It crosses generations.

PARTICIPATION ACROSS SPACE AND TIME

In January 2021, voters in Georgia cast their ballots to decide who would represent them in the US Senate. These run-off elections determined whether the Republicans would retain a majority in the second chamber of Congress. With climate-change denier Donald Trump recently ousted from the White House, Georgia's voters were, in effect, determining the balance of power in the institutions of one of the world's biggest polluters, whose government had, in recent years, made climate injustice much worse. It's hardly an exaggeration to say that the residents of one state, in one rich country, were deciding the future of the world. As it happens, Democrats Jon Ossoff and Raphael Warnock won the seats, putting the Senate at 50:50 and giving Vice-President Kamala Harris the tie-break. But whatever the result, that is not what participatory climate justice looks like.

Climate justice is global.

Groups of academics and activists have been developing a 'World Assembly' or 'Global Assembly' as a first step towards more global deliberation and representation. These groups start small, but have ambitious plans, drawing on 'ordinary people' from across the globe to set priorities or make recommendations at international negotiations.

Climate justice is intersectional.

There will be no justice unless decision-making – local, state *and* global – includes the communities most vulnerable to those decisions. That includes people of colour. It includes Indigenous communities. It includes the global south. It includes displaced communities.

Effective participation means recognizing the overlapping lines drawn by race, gender, nationality, wealth, class, religion and sexual orientation. White women cannot truly represent all women, nor men of colour all people of colour. A supermodel, Black *or* white, may do a good deal to raise awareness – may, in a significant way, act as an ally – but she cannot understand the plight of women of colour clinging to the wreck of their homes. Nor can an Asian-American woman who earns millions in a Wall Street bank.

Climate justice is intergenerational.

As Tuvalu's former prime minister put it, climate negotiators should listen to young activists, not fossil fuel companies. In 2020, 330 young delegates from 142 countries came together in a Mock COP (Conference of Parties). Nearly three-quarters were from the global south. They agreed proposals on climate education, climate justice, climate-resilient livelihoods, physical and mental health, countries' commitments to reduce greenhouse gases, and protecting biodiversity. That could be a start, but, as with citizens assemblies, only if politicians take them seriously.

Given how much today's teenagers have at stake, climate justice almost certainly means lowering the voting age. But what of young children? What of those who have not yet been born?

'Future generations cannot speak,' writes Peter Lawrence, a scholar of international law and philosophy. 'But their interests can be represented.' Future people could be advocated for in national governments, or through a United Nations High Commissioner for Future Generations. This would promote solidarity across generations: support governments, research policy practices, advise on implementing commitments to future individuals' rights. Such an idea was proposed, but rejected, at the Rio+20 conference on sustainable development.

RECOGNITION

In the civil rights movement, Black Americans carried signs saying, 'I am a man.' Today, school strikers' placards declare, 'It's our future.' The Standing Rock Sioux Tribe, protesting against an oil pipeline, is fighting cultural destruction. The Fort Laramie Treaty was supposed to ensure 'undisturbed use and occupation' of reservation lands; this, the Tribe points out, is being violated.

The philosopher Charles Mills observes that liberal states have revolved for centuries around systematic *dis*advantage, based on class, race and gender. Certain groups have been deprived of more than interests, rights, and the chance to participate in decision-making: they have not been seen and acknowledged as equal members of society at all. Philosophers call this 'recognition'. Climate injustice feeds, as we've seen, on systematic *mis*recognition. Those with most at stake are made invisible.

This, says the feminist philosopher Nancy Fraser, is about social and economic culture as well as legal rights. It's a matter of unchallenged, sometimes unacknowledged norms. It's about the dominance of a certain (white, masculine) mindset. It's also about money, about global capitalism. Climate denial, powered by the very rich, has distorted policymaking; much of the media is owned by billionaires. When it comes to resources, *equality* need not be the goal; but so long as some are rich enough to buy political influence, the basic rights of others are too easily set aside.

A JUST TRANSITION

'Coal was the backbone of our community,' the Appalachian miner William Muncy told the *Guardian* in 2020. He was a widower with two kids to support, and he'd just lost his job.

There is climate justice, and there is where we are now. The next chapter will grapple with the apparently unbridgeable gulf between the two. But even a smooth, rapid transition – one that neither 'grandfathers' nor relies on stratospheric technological fixes, and that makes the richest states and corporations pick up the tab – is not without its potential losers. Not everyone connected with those states or corporations *is* rich and, for those who have worked their whole lives in coal, oil or gas, a way of life is also at risk.

Climate justice requires rapid decarbonization. Anything less would be to sanction unimaginable suffering. Even the US Mine Workers Union recognizes as much. In spring 2021, it pledged to support President Biden's shift away from fossil fuels. But it is no justice to throw meat and dairy workers, air stewards and stewardesses, oil and gas employees to the wall. Nor to leave elderly people unable to heat their homes, or the disabled to get about their cities without assistance. It is, on the contrary, fundamentally wrong.

The Stockholm Environment Institute lays out principles for a just transition, in line with this. Active decarbonization (of course). Not investing in carbon-intensive industries. But also supporting affected regions, especially those less able to diversify. Supporting workers, families and the wider community. Climate justice means career change opportunities, retraining, social protections. In return for their support for the shift away from coal, the US Mine Workers Union required measures to help miners, including a pledge to create thousands of renewables jobs. Climate justice means cleaning up environmental damage, not leaving regions scarred. It means inclusive and transparent planning. It means empowering vulnerable groups: those whose basic rights could be threatened by getting mitigation wrong, as well as those with most to lose from *not* mitigating.

WHAT ABOUT NON-HUMANS?

Mitigation, adaptation, compensation; a fair distribution of their costs; equal participation and structural change. And a just transition, to protect those caught in fossil fuel jobs. That's already quite a long 'to do' list. But if Chapter Three was right, it should be longer still.

Humans aren't the only ones who matter. If we owed it to other species and systems not to harm them, we now owe them rectification, for failing in that duty. This makes mitigation even more important. It also greatly expands the demands of just adaptation.

Alvin DuVernay, a Hurricane Katrina survivor, describes taking his fishing boat through floods, rescuing pets as well as people. Adaptation projects, says the philosopher Angie Pepper, should protect non-humans that share our cities, towns and settlements, not violate the rights of wild animals. There must be no barriers to stop wild animals from adapting, if they can. They must, if necessary, be helped to do it.

So far, so good. But, in practice, conflict is inevitable. Projects to redress past damage must be weighed against further injustices that might be done in the process. Some things we can be sure of, from our moral starting points. A policy is not just if it sacrifices vulnerable humans. Conservation colonialism is never just. But what of other conflicts? Restoring ecosystems can mean killing invasive species. Even renewable energy can be bad for wildlife.

Non-humans, like future generations, cannot speak for themselves. But they can be acknowledged. They can be represented.

Since 2014, New Zealand has granted the Te Urewera Park, the Whanganui River and Mount Taranaki the legal status of 'personhood'. In 2010, international social movements and grassroots climate groups at the Cochabamba Climate Conference in Bolivia called for a 'Universal Declaration of the Rights of Mother Earth', and demanded an international climate court. According to scholars of multispecies justice, it's time to revive this idea.

On the face of it, then, acknowledging the moral claims of non-humans makes climate justice harder. But perhaps it also provides us with a way forward. Multispecies justice, remember, puts human–animal interactions in a broader context. It reminds us that we are part of the ecosystems we destroy. The key idea? To learn from different communities and end oppressive relations: those with each other *and* those with the non-human world.

The Iroquois concept of 'seven generations' means taking into account the experience of the last three generations, and the need to support the next three, in addition to our own, when making decisions. The Maori fought over many years for legal protection for the Whanganui River, which they recognize as a living being. 'Collective continuance', says the Indigenous philosopher Kyle Powys Whyte, recognizes reciprocity, within communities and with non-humans. A community has a responsibility to the fish, to the trees; these, in turn, feed and sustain other species. Collective continuance acknowledges our human dependence on the world *and* its potential to flourish in its own right.*

These communities have interacted with our finite world without commodifying it, taking it for granted or exploiting it. If we had listened to them sooner, there would be no need for me to write this book.

*A note of optimism: as I write, Whyte himself serves on President Biden's Environmental Justice Advisory Council.

5

The Least Unjust Option

On 12 December 2015, just as countries were due to adopt the Paris Agreement on climate change, the United States came up with an obstacle. It demanded a weakening of the normative language: developed countries 'should', rather than 'shall', make quantified, economy-wide cuts to carbon emissions. According to the *Guardian*, the US delegation said the original 'shall' was an error, approved by oversight. But for poorer countries, the more demanding word represented a moral imperative: rich, high-polluting states must take the lead.

The US got its way. The change, says European Union delegate Radoslav Dimitrov, was 'slipped in as a "technical correction", together with punctuation changes!'

In the real world, there is no magic wand to bring about the way forward outlined in Chapter Four. Instead, we have a muddle of international negotiations, under pressure from social movements on the one hand, and vested interests on the other.

Too often, rich polluters either refuse outright to do what's right, or make grandiose claims without actually intending to change. This leaves a choice, for those who *do* care. It's crucial to make that choice in the least unjust way, but there's no pretending it's anything other than what it is: a moral second (or third) best.

That's what this chapter is about.

THE PARIS COMPROMISE

The Paris Agreement on climate change was adopted on 12 December 2015. It was a historic achievement: a step many had come to believe was impossible. It was the culmination of more than two decades of international negotiations and innumerable long-haul flights and late-night bargaining sessions. It is the best we currently have.

Unfortunately, it's a long way short of enough.

Whatever participatory justice requires – transparency, recognition, participation on equal terms – it looks very different from the process that actually produced the agreement. The agreement was the result of climate diplomacy, mostly conducted behind closed doors. Indeed, it was widely considered an unexpected victory of just such diplomacy that any agreement was reached at all, when the parties had such different agendas. Big polluters generally didn't want binding agreements on emissions cuts; rich countries often wanted to ignore global adaptation needs. Small island states and more vulnerable countries wanted ambition and compensation. It took clever leadership by the French hosts to reconcile them into a deal.

Radoslav Dimitrov, a scholar of environmental politics as well as a climate negotiator, paints a picture worthy of the final scenes in a suspense movie. Secret negotiations, alliances forming across groups, many delegates unaware of what had been settled until they were presented with a *fait accompli*. The result? Small island states gained some points, mainly via a 'high ambition' coalition that also included the European Union. But mainly the result reflected the balance of power going into the negotiations. The developed countries of the north, says Dimitrov, 'won most of the key battles'.

That doesn't bode well for climate justice. But let's take a closer look.

WORDS WITHOUT COMMITMENTS

On the plus side, the Paris Agreement *is* a global deal. And it does, to some extent, acknowledge the demands of justice. Here it is, highlighting divergent degrees of responsibility for harm and capacity to act.

> In pursuit of the objective of the [UNFCCC], and being guided by its principles, including the principle of equity and common but differentiated responsibilities and respective capabilities, in the light of different national circumstance . . .

There are promising words on protecting those in the greatest need, and upholding human rights, including women's rights:

> Also recognizing the specific needs and special circumstances of developing country Parties, especially those that are particularly vulnerable to the adverse effects of climate change . . .
>
> Acknowledging that climate change is a common concern of humankind, Parties should, when taking action to address climate change, respect, promote and consider their respective obligations on human rights, the right to health, the rights of Indigenous peoples, local communities, migrants, children, persons with disabilities and

> people in vulnerable situations and the right to development, as well as gender equality, empowerment of women and intergenerational equity . . .

There's a nod to a just transition:

> Taking into account the imperatives of a just transition of the workforce and the creation of decent work and quality jobs in accordance with nationally defined development priorities . . .

Non-human justice is mentioned, however passingly and indirectly:

> Noting the importance of ensuring the integrity of all ecosystems, including oceans, and the protection of biodiversity, recognized by some cultures as Mother Earth, and noting the importance for some of the concept of 'climate justice', when taking action to address climate change . . .

There's also a legally binding component: a procedural requirement for states to come up with increasingly ambitious commitments to reduce greenhouse gas emissions every five years. These are known as nationally determined contributions (NDCs). States must also report on whether these have been met, and take part in a 'facilitative, multilateral' consideration of overall progress. Failure to set ambitious mitigation targets, *or* to stick to them, could damage reputations. In theory, this might be enough to scare major polluters into action, at least if other states have been more ambitious. But it's a big 'if' and a big 'might'.

Now the bad news. The quotes above sound promising, but they're all from the Preamble, not the actual agreement text. What counts, legally speaking, is the operational material, and then it depends on the precise language. And this, unfortunately, is where it gets depressing, because much is left voluntary or imprecise. NDCs are just that: *nationally* determined. As the legal experts Lavanya Rajamani and Daniel Bodansky point out, states might be told to determine their targets in line with 'equity' and 'common but differentiated responsibilities', but they choose how to apply these terms.

Rajamani picks the agreement apart, distinguishing 'hard' or legally binding obligations from 'soft' ones, or 'mentions' that don't count as either. Most obligations around adaptation – addressing loss and damage, financial support, technology and capacity-building – are soft. Take Article 8, paragraph 1:

> Parties recognize the importance of averting, minimizing and addressing loss and damage associated with the adverse effects of climate change, including extreme weather events and slow onset events, and the role of sustainable development in reducing the risk of loss and damage.

Encouraging words, but without the mandatory requirements to turn it into anything like justice. Indeed, in adopting the agreement, the negotiators made explicit that Article 8 'does not involve or provide a basis for any liability or compensation'. Small comfort for those disappearing small island states.

Ultimately, the problems with the Paris Agreement can be summed up in a few words, quoted above: 'Noting the importance for some of the concept of "climate justice" . . .' As we've seen, climate justice isn't an esoteric idea, to be weakened with quote marks. It's not a controversial goal we should acknowledge because some people happen to value it. It's basic morality. It should be important to all of us, and integral to climate action.

WE DON'T EVEN HAVE PARIS

We'll always have Paris. Except we won't, as we saw when President Trump took office in 2016, and the US announced its withdrawal from the agreement. As I write, America is back in, under the more future-friendly Biden administration. This is encouraging. But it also shows how entirely the success of that long-fought step in the right direction relies on the domestic politics of a few big players.

In September 2020, China announced that it would be carbon neutral by 2060. By December 2020, President Biden had proposed a 2050 target for the US. In November 2021, world leaders met in Glasgow for their *twenty-sixth* Conference of Parties (COP). Afterwards, independent scientific analysis by Climate Action Tracker concluded that the submitted, binding long-term net-zero targets and 2030 NDCs could keep global warming down to 2.1°C by 2100, if they were fully implemented. Including all announced (but not yet binding) targets, it would be 1.8°C.

That's something, but it's still well above 1.5°C. And the difference is crucial. It means more flooding, more extreme weather, more heatwaves. It means more people getting sick, losing homes and livelihoods. More people dying.

What's more, full implementation looks unlikely. The 2020 UN 'Emissions Gap Report' was not encouraging. According to Climate Action Tracker, based on only 2030 NDC targets, the planet will be 2.4°C hotter than pre-industrial levels by the end of the century.

Moreover, pledges must be matched by action. Based on national policies adopted by November 2021, temperatures are set to increase by 2.7°C. That's terrifying territory. In May 2021, the International Energy Agency said *all* new fossil fuel projects must cease that year, to reach net zero by 2050. But the Glasgow Climate Pact (as it is known), was woefully weak on fossil fuels, only asking parties to 'accelerat[e] efforts towards the phasedown of unabated coal power and phase out of inefficient fossil fuel subsidies'.

If you think this failure to mitigate is because everything is being flung at adaptation and compensation, think again. Advanced economies promised $100 billion a year for a Green Climate Fund by 2020; This promise had not been kept by the 2021 COP, although a report spearheaded by Germany and Canada claimed that the full amount would be raised by 2023. And adaptation, where it *is* happening, isn't necessarily just adaptation. Remember those Australian councils' plans, at odds with what communities wanted? That's not unusual. A common approach, according to David Schlosberg, is to go heavy on the big picture – government-level risk assessment, legal liability, infrastructure and emergency response – and neglect the community-led planning actually needed to protect the vulnerable.

Put non-human justice in the mix, and things look even worse. We're in the middle of the sixth mass extinction. By 2020, 515 mammal, bird, reptile and amphibian species were on the brink of disappearing, according to ecologist Gerardo Ceballos and colleagues. Between 1900 and 2019, 543 had already died out. By 2050, the extinction rate is likely to be 117 times higher than the normal rate for the past two million years. The Living Planet Index, which measures diversity, dropped 68 per cent in the 45 years to 2016.

'An SOS for Nature', in the words of the WWF. But it's an SOS for humans, too. This mass extinction, says Ceballo, 'may be the most serious environmental threat to the persistence of civilization'. Because it's irreversible.

GREENWASH AND BULLSHIT

'It's goodbye climate deniers,' writes journalist Damian Carrington, 'hello climate bullshitters.' His bluntness is justified. If there was a word for the early 2020s (apart from COVID-19), it should be 'greenwashing'.

Some policies, Carrington points out, push emissions *up*. New roads, airport expansion, ditching home energy efficiency programmes. As I write this, the UK government is dithering over a coal mine that should have been decisively blocked, and looks set to approve a huge new oilfield west of Shetland. Across the globe, states pour cash into subsidizing fossil fuels – $180 billion in 2019, on International Energy Agency figures – and ignore the climate impact of animal agriculture.

In 2020, a group of environmental organizations produced the aptly named 'Production Gap Report'. It found that fossil fuel production must fall by 6 per cent yearly from 2020 to 2030, to keep temperature rises to 1.5°C. As of 2020, countries were planning on *increases* of 2 per cent a year.

In terms of greenhouse gas emissions from livestock, Oliver Lazarus and colleagues did the sums on the business-as-usual emissions of Nestlé and dairy giant Fonterra. The verdict? They would more than use up the total 2030 emissions targets of their countries, Switzerland and New Zealand. Yet neither country, Lazarus says, even mentions animal livestock or animal agriculture in its climate commitments.

As for the corporations themselves, there seems to have been a shift towards greener messaging, but little else. In 2020, the Advertising Standards Authority banned, as misleading, an advertisement describing Ryanair as 'Europe's lowest fares, lowest emissions airline'. The same year, the Dutch Advertising Code Committee ordered Dutch airline KLM to change a campaign that over-emphasized the firm's use of biofuels.

The ten biggest US meat and dairy companies have, say Lazarus, 'lacked transparency about their emissions, lacked sufficient mitigation targets, or worked to influence public opinion on climate policy'. Glossy references to 'sustainability' notwithstanding, they're not targeting the real source of greenhouse gases. They might switch to renewable energy, but they won't get shot of the farting cows.

Oil and gas firms run advertisements acknowledging climate change, touting the benefits of technology or citing net zero aims. But they keep on extracting fossil fuels. And banks go on financing them. To the tune of $3.8 trillion between 2016 and 2020, according to one NGO report. And that's just from the sixty biggest commercial and investment banks. So-called 'negative emissions', via carbon removal, are needed to counterbalance genuinely hard-to-reduce emissions. Renewables are crucial to produce energy *without* burning fossil fuels. But, warns Simon Lewis, professor of global change science, some big finance organizations and fossil fuels seem to have forgotten this. They treat renewables or carbon removal as offsets to justify 'business-as-nearly-usual'.

Nor does the media help. Fossil fuel advertising skews viewers' impressions of the climate emergency; we saw that in Chapter One. But it goes deeper than that, too. In 2020, Silvia Pianta and Matthew Sisco examined a dataset of 1.7 million news articles across the EU, from 2014 to 2019. They found that media coverage of climate change mostly goes up when there are short-term weather changes, not when scientists highlight long-term deviations from climate norms. Even progressive newspapers continue to advertise petrol cars, sell far-flung holidays, and feature meat and dairy recipes (although the *Guardian* has at least rejected fossil fuel advertisements). Right-wing news outlets are only now pulling back from full-blooded climate denial, and even here there are exceptions.

WHERE FROM HERE?

The last chapter asked who should pay for mitigation, adaptation and compensation. Now, we face the reality: many of those who should pay won't. So how should everyone else respond? Philosophers call this 'non-ideal theory': figuring out what to do in a world where circumstances are very far from favourable, and major players (to put it mildly) will not do their bit.

The next chapter will talk individuals. For now, let's ask what institutions should do. How should other states have reacted had the Trump administration remained in power, with the US outside the Paris Agreement? How should they respond to Russia, whose NDC is, in Henry Shue's words, 'farcical'? How should governments treat corporate Goliaths who go on hunting for gas and oil, and draw on their vast resources to fight responsibility for devastation through the courts? How should councils and politicians react if wealthy, able-bodied citizens choose cars and planes over bikes and trains?

The best option, Simon Caney points out, is to require those who should pay to fulfil their obligations (or pass the costs onto them if they don't). Within states, this might be through better public transport or changing urban design. It might be through carbon taxes, tagging the true climate cost onto the price tab for high-carbon activities, *so long as the vulnerable are protected*. The caveat is key: if a carbon tax leaves the very poor unable to heat their homes or get to work, it perpetuates one fundamental injustice in seeking to alleviate another.

A just tax could apply a differential rate, reserving the heftiest surcharges for what Columbia University's Philippe Benoit calls 'discretionary extravagant activities', from space tourism to first-class air travel or petrol sports cars. It could earmark a proportion of revenues to keep energy affordable for the most vulnerable. It could help fund the shift to renewable energy.

When the non-compliers are state governments themselves, the options are more limited – and more difficult. Trade sanctions. Continued international diplomacy. Ultimately, reforming the global institutions that have proved so adept at promoting the status quo.

And failing this? Morally, there's a straightforward answer. Other wealthy players should pick up the slack. If some who should don't, others may have to make quicker, sharper emissions cuts, invest more in renewables and spend more on adaptation aid. It's the next best, the 'least unjust' option. They can afford it.

There's an objection, of course: that timeless plea of the young child, 'It isn't fair!' And it isn't. It isn't fair on those who have done their 'bit' to expect them to do more. That's the same whether we're talking about tackling climate change or helping out in your kids' school. And such unfairness *is* morally problematic. We saw that already, in the intuition against free riding, in the influential idea of social justice as fair cooperation.

But what's the alternative? *Not* picking up the slack isn't a morally neutral option. The philosophers Dominic Roser and Sabine Hohl spell this out. It amounts to failing to protect human rights: to do even basic justice. If no one else pays the tab, the victims of climate change will pay it, in slow deaths from malaria and fast ones under tsunami waves. In starvation, in forced migration, in losing the things and places they hold most dear.

The bottom line? Unfairness matters, but basic rights matter more.

But, in practice, even this looks optimistic.

WHO *REALLY* PICKS UP THE SLACK?

There is some hopeful news.

According to a Zero Carbon Campaign poll in 2021, two-thirds of people in the UK think a carbon tax would be a fair way to fund a green recovery from the coronavirus pandemic. Of the 2000 polled, 68 per cent also wanted poorer people protected from the impact of the taxes. Four out of five members of the Irish Climate Assembly were willing to pay higher taxes on carbon-intensive activities, so long as the poorest citizens were shielded and revenue used to transition to a low-carbon, climate-resilient economy.

Meanwhile, according to the thinktank Carbon Tracker Initiative, seven 'virtuous' feedback loops are driving energy systems away from fossil fuels: from a positive cycle of growing demand and falling costs, to geopolitics.

But governments' failure to end fossil fuel subsidies – clearly the absolute least they should do – remains a depressing indictment of where we are now. Both Paris and Glasgow made horribly plain the limits of international diplomacy. There are political difficulties around trade sanctions, especially if some of the most powerful states in the world would be at the receiving end.

In practice, the obvious 'next best' option is also problematic. For one thing, when it comes to mitigation, it's not certain that even powerful states *could* 'pick up the slack' in the face of others determined to blow the carbon budget. As Stephen

Gardiner points out, a state or two, burning enough coal and oil, could do that all on their own.

For another, it's extremely unlikely that any rich states *would* take up others' responsibilities, when they are so disinclined to fulfil their own. As of September 2021, The Gambia was the only state analysed by Climate Action Tracker whose NDC *was* compatible with the 1.5°C target. Chapter Four began with an American president claiming it was unfair for the US to do what, on any reasonable philosophical account, is *less* than its fair share.

So who really has to pick up the slack? Future generations? That's one possibility: cutting emissions now but passing the economic costs on (essentially, borrowing from the future to pay current generations to make the changes). The philosopher John Broome suggests this, but acknowledges the unfairness. Already vulnerable people? Desert Sunlight, a vast solar plant in the Californian desert, is a victory for mitigation. But for the Colorado River Indian Tribes it is a violation of sacred lands. Or non-humans? Lithium mining for electric cars wrecks the biodiversity of Andean salt lakes. Tidal barrages can destroy wildlife.

None of these are options we should feel comfortable about.

THE BOTTOM LINE

This is the situation we are in: a world of tough trade-offs, forced on everyone else because powerful players won't do what they should. It's horrible. But it's no reason to give up on justice altogether, to grab at whatever looks politically easiest. Moral guidelines are essential, to navigate the terrain of far-from-ideal justice.

Here are two such.

First, not everything should be on the table. Choosing between lesser evils, we need a 'don't cross' line. Basic justice is that line; human rights are that line. If uranium mining gives the Navajo people cancer, it is not – and cannot be – part of climate justice. Land grabs for biofuels could push millions into starvation. Or so C. Ford Runge and Benjamin Senauer warned, back in 2007. That's not justice. It's fundamentally immoral.

That's one compelling criterion. But it doesn't make everything straightforward.

Consider, for example, sacrificing things that some people consider important but aren't essential to basic justice. At one level, this is a morally compelling way forward. If someone like me, who loves hiking, is squeamish about cluttering hills with wind turbines, that's no reason to perpetuate rights violations or force others to sacrifice almost everything. It seems, rather, a self-indulgent version of 'Nimbyism'.

However, not all values or ideals are equal. Some look more like preferences, or 'moral luxuries'. Others are culturally central: themselves a matter for basic justice. Exactly where – and how – should *that* line be drawn? Too often, it is the values that have been treated as expendable for centuries that are sacrificed again in the name of climate action. Reducing emissions from deforestation and forest degradation ought to be a win-win situation, cutting greenhouse gases and protecting Indigenous communities. It's not, warns the lawyer Annalisa Savaresi, if it ends up excluding communities from the places that matter most to them.

Mitigation colonialism? That's not justice either.

What's more, some of these 'values' are views *about* non-humans. Their fate matters – at least arguably – for their own sake, too. The desert tortoise, burrowing deep into the Mojave desert, loses its habitat to solar panels. This, too, must be weighed in the balance.

That's where the second point comes in. It's more vital than ever that vulnerable groups are recognized, that their voice is heard, in making these morally difficult decisions. That they have a seat at the table.

We talked in Chapter Four about climate assemblies, about citizens being informed, deliberating, and being heard; about global decision-making by all, not only the most powerful states. We talked about the Cocobamba Climate Conference, proposing rights for 'Mother Earth'. We talked about a representative for future generations.

Of course, talking is easy. Citizens' assemblies need recognition by national and international powers, to have real influence. The signals from Ireland, after its 2016 to 2018 Climate Assembly, are mixed. The minority government stifled a bill in 2018 that might have given teeth to citizens' recommendations, including by banning new oil and gas exploration licences. A Climate Action Plan was passed, but, according to one civil servant interviewed by researcher Louise Michelle Fitzgerald, it 'discard[ed] the citizens' recommendations'. Environmental groups were similarly unimpressed.

Despite this, Fitzgerald thinks the anti-fossil-fuel norms in Ireland will have lasting implications. In 2021, the government did finally ban new oil and gas exploration licences. Citizen-led, grassroots climate politics may yet be taken seriously.

But for true participatory justice, the decision-making (and the representation) needs to be global and intergenerational. In practice, the High Commissioner for Future Generations was rejected by world leaders. Stephen Gardiner, despairing of progress through the UNFCCC negotiations, calls for a 'constitutional convention' on the climate crisis. The idea? For parties to come together and design a new global constitutional system that actually protects future generations.

One thing is clear: something needs to be made to work, and to work globally. Not everyone is motivated by climate justice. That's the grim reality. But for those who are, this is the choice. Recognize the interdependence of humans and 'nature', the history of intersectional exploitation, the plight of future generations. Make the vulnerable visible. *Or* try to separate climate action from climate justice, make inevitable trade-offs on short-term economic or political grounds. And watch history repeat itself.

BUILDING A BETTER WORLD

A Joel Pett cartoon of a climate summit regularly does the rounds on social media. Aims on a board:

'energy independence', 'preserve rainforests', 'healthy children'. A staring audience. Alongside it, in a speech bubble, are the words: 'What if it's all a big hoax and we create a better world for nothing?'

It's not a hoax. But what indeed? That's one more thing to hold on to: what getting this *right* is really about.

The 2020 World Happiness Report, combining Gallup World Poll data from 2009 to 2011, said 62 per cent of respondents would prioritize environmental protection over economic growth. But in 2021, Swiss voters narrowly rejected key climate measures in a referendum. And why? 'Climate action was framed as economic sacrifice, even by its advocates,' says Julia Steinberger, professor of social ecology and ecological economics. 'Acting to preserve living conditions for the future . . . was presented through the lens of economic inconvenience, instead of an urgent core priority of any self-respecting citizen.'

There are, says Simon Caney, some obvious ways forward, in the maze of non-compliance. Build electoral support for mitigation, for one. Also, present a vision of a low-carbon future. In other words, communication, communication, communication.

Because climate action isn't all about burdens. It is – or should be – about creating green jobs and green spaces, rewarding the caring roles that really make lives better. It's about active and engaged citizens, diets that are healthy, not destructive. It's about healthcare, education and reproductive rights. Green New Deals, proposed on both sides of the Atlantic, would go a long way towards that. Women's rights and climate justice organizations demand a *Feminist* Green New Deal. That would go further still.

Climate justice is about learning from the past. But it's also about looking forward. It's about creating a society – and a world – that people can thrive in.

6

But What Can *I* Do?

As an individual, it's easy to feel powerless. I cycle across the city, planes overhead, cars sliding along the bypass below. I'm vegan, but restaurant menus are stuffed with meat. I haven't flown since 2018, and not for pleasure since 2014, but my social media feed is full of excited plans for the next overseas holiday or business trip. And that's in my climate-conscious echo chamber.

The thing is, I understand. This isn't about casting blame. Naturally, we ask ourselves: 'If everyone else is doing it, why shouldn't I?' What's more, we make decisions within parameters set by governments, corporations, a billionaire-funded media. Why pay a fortune for trains and taxis (even if you can) when you could go door to door by car? How much harder to quit animal products when those around you are enjoying their steaks?

As we've seen, climate injustice is global. It's collective, systematic and institutionalized. That's not the impression given by big corporations, politicians or much of the media, from the Canadian politician who allegedly taunted climate protestors, asking 'how they charged their iPhones', to fossil fuel companies using advertisements to shift responsibility to consumers. Nonetheless, it's true.

World carbon dioxide emissions in 2019 were 36.42 billion tonnes, on Oxford Martin School figures. They dipped in 2020, but that was because of the pandemic. Remember that the twenty biggest oil, natural gas and coal companies caused 493 billion tonnes of carbon dioxide and methane emissions between 1965 and 2018 (35 per cent of the global total). Per person, we produce only 17.74 or 6.64 tonnes annually, even in the high-polluting US or UK respectively. (Or we did in 2019.) That's just a drop in the overflowing ocean.

But, as we've also seen, we start where we are now. That means asking what this collective injustice means for individuals: what we *should* be doing. Especially those who, like me, enjoy a comfortable life in the global north. Because this is an institutional injustice, can we put our feet up, consciences clear, and wait for governments to sort it out?

Absolutely not.

FACING UP TO PRIVILEGE AND COMPLICITY

Climate injustice is systematic violation of human rights. Chapters One and Two made that clear. We take part in patterns of action, we support institutions, which kill people. I put it brutally because it *is* brutal.

Think about it smaller scale. Suppose you belong to a golf club that won't let women of colour join but pays them a pittance to scrub the toilets. Or you have shares in a firm testing dangerous chemicals on children. Can you really close your eyes? With climate change, the structures are more nebulous, containing multiple entities, but the bottom line is the same. We benefit from institutions that flood Bangladesh, drowning children and forcing young girls into prostitution or child marriage; institutions that will rob the citizens of Tuvalu and Kiribati of the lands they call home.

That doesn't make this responsibility *equal.* Anything but. Recall those Oxfam figures: more than half of carbon emissions between 1990 and 2015 were caused by the richest 10 per cent of people. Frequent flyers are only 1 per cent of the world's population, but they rack up half the carbon emissions from commercial aviation. In 2018, only 2 to 4 per cent of people flew internationally at all, on figures from social scientists Stefan Gössling and Andreas Humpe.

But nobody living comfortably in a society made prosperous by fossil fuels can legitimately ignore climate injustice. In fact, if we accept the second basic moral rule I began with, nobody living a comfortable life, full stop, should ignore climate injustice. Each of us should save other humans from severe suffering, if we can do it comparatively easily. As we saw, it's impossible to do this alone, but we can do it together. It is our governments' job to respect human lives, wherever and whenever they are, *because* it is part of what we owe to one another, as human beings.

So no, this isn't – and must never be – about blaming ordinary individuals for climate change. Should we even feel guilt for something we didn't cause on our own, and can only ever be part of redressing? Perhaps, as the philosopher Larry May suggests, we should instead feel shame, because we became who we are, and can live as we do, on the back of such suffering. But we *should* feel guilty for not doing what we can to change things.

WHAT WE CAN DO

As individuals, we should promote social and institutional change. Climate change is a collective problem; it needs a collective solution. Anything else is inefficient, unfair and ineffective.

Suppose, for the sake of argument, that enough people voluntarily 'greening' their lives could keep global warming below 1.5°C. This would be inefficient. It's much harder and more expensive for everyone to change their way of life now than it would be with a few well-chosen infrastructure changes: reliable public transport, affordable and delicious plant-based food. According to the effective altruism charity, Founders Pledge, giving $1,000 to the most effective carbon-cutting charities could save up to 100 tonnes of carbon dioxide. That's up to 25 times more than any of the lifestyle changes they consider.

If change is entirely driven by individual choices, it leaves the costs to fall on those who are actually prepared to do something, rather than distributing them along the lines laid out in Chapter Four. That's unfair. Of course, it's better than not mitigating at all. But it would be fairer if governments ended fossil fuel subsidies, invested collectively in renewables, and spread the load to those who caused the harm and can afford to bear it.

All this is true. It is also, to some extent, academic, because of the final point: effectiveness. The 'sake of the argument' assumption I just made is false.

Individual emissions cuts couldn't be enough for climate justice.

Suppose lots of well-off people cut their carbon footprint. Will it mitigate climate change? That depends on how many of the richest 10 per cent are among them. It also depends on how everyone else responds. Do they follow suit, or do they consume *more* because fossil fuel purveyors drop prices? That's the so-called rebound effect, and we ignore it at our peril. In short, carbon-cutting matters, but it won't work without supply-side constraints.

Moreover, climate justice requires more than mitigation. It means protecting lives. Reasonably affluent individuals can drive less, eat more plants and less beef, buy solar panels or air-source heat pumps. But we cannot, for the most part, give renewable or adaptation technology to the global poor. That takes collective action.

Finally, as we saw in Chapter Two, climate change is not a stand-alone problem. It's no coincidence that the victims of climate change are disproportionately poor, disproportionately women of colour. Without addressing oppression and exploitation, mitigation and adaptation efforts are a sticking plaster, at best, over a constantly reopening wound. That, too, takes collective change.

Ultimately, we need to create (and maintain) a just world. To do that, we must think of ourselves not only as individuals. To borrow from Sherilyn MacGregor, an ecofeminist political thinker, we are also *citizens* in the broadest sense: members of a global eco-political community.

COLLECTIVE ACTION WORKS

Individuals can give money. We can donate to climate justice charities and NGOs. As we just heard, that's a lot better than nothing. But individuals can do more than that to challenge social norms, institutional behaviour and government policy.

We can invest in renewables. We can petition pension funds and banks to divest from fossil fuels. We can support climate education and campaign to ban fossil fuel propaganda. Perhaps we cannot all be Mary Annaïse Heglar, George Monbiot or Naomi Klein, but we can read and circulate what they say. We can sign online petitions, email politicians. We can do what Heglar calls 'greentrolling': mocking fossil fuel giants for hypocrisy on their social media feeds, then sharing it far and wide.

We can run for office. Or, if not – if we cannot emulate Caroline Lucas of the Green Party or

Alexandria Ocasio-Cortez, the progressive congresswoman pioneering a Green New Deal – we can support them. We can campaign. We can knock on doors and explain policies. If we are old enough, we can vote.

We can litigate. In 2015, 21 young people, together with the intergenerational environmental organization Earth Guardians, filed a lawsuit against the US Government. *Juliana v. United States* accuses the government of violating 'the youngest generation's constitutional rights to life, liberty, and property' by causing climate change. Elsewhere, climate-conscious children from Portugal to Peru are also turning to legal action.

We can protest. We can march in the streets, picket banks, insurance companies, government departments, and the offices of Shell, BP and ExxonMobil. We can engage in civil disobedience. In August 2018, a schoolgirl stood outside the Swedish parliament, holding a sign saying 'Skolstrejk för klimatet' ('School Strike for Climate'). Her name, of course, was Greta Thunberg, and she started a global movement. Before the pandemic hit, millions joined her. On 6 November 2021, around 100,000 marched in Glasgow.

But does any of this work? There, of course, is the rub. Thunberg's book is called *No One Is Too Small to*

Make a Difference, but in 2019 she told activists that the global strikes had 'achieved nothing'. Greenhouse gas emissions were still rising.

In fact, when it comes to the climate strikes, it's too early (and too complicated) to tell. Sociologists Dana Fisher and Sohana Nasrin tell us the peer-reviewed evidence isn't there yet. But if nothing else, Fisher points out, the strikes have created a cohort of active citizens, future voters who know the facts about climate change, and care. That is – or should be – a great deal. To condemn thousands of school strikers for missing out on their education, as some Conservative politicians have, looks decidedly short-sighted. Ultimately, political parties cannot afford to alienate the next generation of voters.

What's more, that's almost certainly *not* all the strikes have done.

Activism can make a difference. Fisher and Nasrim confirm that. From lobbying to divestment campaigns, to community protests against specific projects. According to the non-profit Effective Activist, success takes a combination of strategies: so-called mainstream tactics (voting, lobbying, drafting legislation) *and* non-violent protest. Boycotts, civil disobedience, strikes, sit-ins, rallies. They're all part of the mix.

Public opinion matters for environmental policy change, and there are ways to change public opinion. One is non-violent activism. Psychologists Belinda Xie and Ben Newell think climate striking can 'promote the psychological factors most important for fighting against climate change'. The sociologist Dylan Budgen found that peaceful climate protests help get the public onside, at least with Democrat and Independent voters.

Another way is targeted advertising. The 2019 New Climate Voices campaign featured people handpicked to resonate with conservatives. Retired Air Force General, Ron Keys; climate scientist and evangelical Christian, Katharine Hayhoe; former Republican Representative Bob Inglis. It targeted two competitive congressional districts for a month. The result? According to Yale's Matthew Goldberg and other communication experts, the views of Republicans shifted several percentage points.

CHANGING HOW YOU LIVE

Does this mean we should pressure institutions, from local governments to the UN, but otherwise carry on exactly as we are? The philosopher Walter Sinnott-Armstrong suggests it's alright to drive an SUV for fun while campaigning for it to be illegal to do just that. Is that right?

No. First, changing your lifestyle is a *way* to promote social and political change. It shows you're willing to accept legislative or corporate adaptations. It can also influence how others behave. In 2013, sociologists used research accomplices to demonstrate composting in a university cafeteria. Others started to copy them. In another study, New Zealand high school students were more likely to recycle if their friends had done so.

I'm not saying you (or I) can do this alone. Perhaps, if you're Billie Eilish or Lewis Hamilton, your influence is such that you can prompt thousands to go vegan just by doing it yourself. Most of us couldn't. But we can do it *with others*. The vegan or flexitarian movement prompts supermarkets and restaurants to offer more vegan food, which makes it easier for more people to cut out meat and dairy. Me quietly quitting flying probably won't get noticed. But by May 2021, more than 4,000 people had signed up to Flight Free UK 2021. That might.

What you do as an individual and what happens politically are not two separate spheres. Whether it's campaigning for change or forging new lifestyle norms, we do what we do as *part* of civil society. We operate within families, faith groups, local groups and NGOs; the many ways we group ourselves and influence each other beyond political or business relationships.

Second, driving, flying, eating dairy and meat aren't morally neutral. Driving that SUV makes you part of the harm: an active contributor to a system that imposes huge suffering on other people. As Chapter One discussed, it makes you complicit.

'BUT DO *I* MAKE A DIFFERENCE?'

Here's an objection. Complicit or otherwise, each of us is only one person. It's all very well talking about citizenship, civil society, and 'what I can achieve with others', but what if those others could achieve it without me?

I have just said that I (or you) will probably make no difference to the harm done by climate change, by giving up flying, meat or dairy. The tornados, cyclones and wildfires will almost certainly happen either way. The same is true if I go to a protest, one among thousands. Why, then, should I bother? After all, it's not cost-free for me. I'd rather not drag myself and my kids to a climate march, on a rare sunny day. It's inconvenient to get two buses rather than drive. I still miss deep-fried Camembert after years as a vegan. I'd rather buy gifts for my family than donate to the Sunrise Movement.

This is a philosophical puzzle, so let's drill down a bit.

First, there's a difference between 'almost certainly won't' and 'won't'. Don't, says Derek Parfit, ignore the small chances of making a big difference. Sometimes that difference can be very big indeed. Think again of the voters in Georgia, January 2021, or any of the swing states in recent US elections. The chance of one vote swinging the result is incredibly tiny, but if it came to be, what implications for America, and for the world!

In protests, numbers also matter. According to political scientists Erica Chenoweth and Maria Stephan, if 3.5 per cent of the population is actively involved in a peaceful movement for change, it should succeed. Thinking like Parfit, in turning up, you have a tiny chance of being pivotal. Suppose Thunberg hadn't bothered, because her solo school strike would probably go unnoticed. Would the millions-strong Fridays for Future movement be what it is now?

The same, philosophers argue, could apply to our individual carbon footprints. You probably won't kill anyone through climate change by flying to the Bahamas on holiday. But total climate harm gets worse, broadly speaking, as total emissions go up. At some point, increasing greenhouse gases *must* make a difference. A cyclone will be worse; the ice cap will melt faster. Thinking like this, your long-haul flights probably won't tip that threshold. But if they do, the consequences are terrible. Put more formally, if there's a small probability that your action will cause extreme harm, then there is an *expected* harm associated with deciding to do it. That's moral reason, in itself, not to do it.

But even if this weren't true – even if, in practice, no one person's actions could be pivotal in this way – it can still matter what you do.

In 2019, Belinda Xie and her colleagues surveyed 921 Australians to find out what swayed them to cut their carbon footprint, support social intervention on climate change mitigation, engage in climate advocacy – or not do so. One of the biggest factors was how ineffective they felt, and this, Xie later elaborated, is alleviated by the 'aggregation effect': being part of a group who can make a difference between them.

We can make philosophical sense of this.

Picture a field, after a battle perhaps. A hundred people, desperately thirsty; a hundred others, including you, each with water to spare. The water is collected in one big container, then handed out to the thirsty men. If you donate your supply, the difference will be so tiny that no dehydrated man will feel any better. But if you all yield your water, they will all be much better. So why should you do it?

Ask yourself, says the philosopher Julia Nefsky, whether the *group* of people still deciding what to do, of which you are one, can make a difference. If you know it can't, giving up your water would be futile. (Suppose the container is already full when you get there, or there is not enough water, even between you all, to help anyone.) But that's not the case here. You can choose to be what Nefsky calls a 'non-superfluous part of the cause' of saving the men. Put more straightforwardly, you can *help* to do it.

There's another moral reason for chipping in, too: fairness.

Fairness matters, intuitively, and in liberal justice. We've seen that already. It matters that you do your bit in the cooperative schemes from which you benefit. The same is true, says the philosopher Garrett Cullity, of shared moral ends. If we should end basic injustice between us – and I argued in Chapter One that we should – it's unfair to leave all the work to others. If they're campaigning for climate justice or cutting emissions between them, you owe it *to them* to do the same. Otherwise, you're free riding.

WHAT TO PRIORITIZE

How far should we take what I've just said? If our individual carbon emissions matter, because they're unfair, they help to harm, or they just might trigger disaster, must we unwind what I said before, about collective action taking priority? After all, a starting point for this book is that it's wrong to harm others. Does this mean we simply shouldn't do *anything* that comes with a carbon footprint, full stop?

Some think so. But that's not the takeaway from the section above. For one thing, imagine the consequences of an all-powerful, super-stringent individual duty *never* to help to harm, or to run even a tiny risk of causing harm. You could do almost nothing. Never travel, even by bike, even to get to a climate protest, in case you cause an accident. Never cross a road. Never eat anything in a public place, in case it triggers someone's allergy. For another, doing your 'fair share' isn't just about carbon emissions. It's about doing your bit, overall, to challenge climate injustice.

The takeaway, instead, is this. Thinking *as* individuals, we can make sense of the injunction to think *not* as individuals: to focus on what can be achieved collectively. That applies to our own carbon footprints, but also, especially, to changing the structures and societies of which we are a part.

Let's be clear. This isn't a claim about offsetting, or 'cancelling out' your contribution to harm by doing climate-friendly things. I'm *not* saying that so long as you give to the Clean Air Task Force or ClientEarth, plant enough trees, or go to enough protests, you can zip around in your SUV with a clear conscience. If you can cut your footprint *and* campaign for climate justice, you should. And, often, carbon-cutting is perfectly compatible with activism. As we've seen, it's part of it.

But the two goals can clash. Some people, on occasion, might do more for climate justice if they increase their own carbon footprint in the process. Consider the so-called 'Al Gore effect', flying round the world to promote political change. (But remember, too, that Thunberg makes a point of *not* flying.) More generally, there are limits to what each of us can reasonably do. You might not have the money to travel by train rather than drive and donate to a climate justice charity. You cannot spend hours researching solar panels and spend those same hours petitioning your pension fund to quit fossil fuels.

The short version? Yes, cutting your carbon footprint matters. But so do a lot of other things. And we shouldn't, in the process, lose sight of the bigger picture. Painful as it is, being an activist might mean learning to live with *some* ongoing complicity, so you can focus on changing the system.

HOW TO DECIDE: CARBON COUNTING

What you cut matters. It's not the aim of this book to tell you the footprint of everything: whether coffee is worse than chocolate, or exactly how bad beef is. That information is out there already. But the recommendations of policymakers and educators aren't necessarily in line with what cuts emissions the most. This information gap matters. So let's talk about it.

In 2020, John Halstead and Johannes Ackva calculated emissions savings from individual lifestyle choices. They're shown in Table 2.

TABLE 2: *Individual carbon-cutting*

Lifestyle change	*Tonnes of CO2 avoided per year*
One child fewer than would have had	4
Live car-free	2.4
Switch to electric car	2
Avoid a Transatlantic flight per year	1.6
Switch to green heating	0.8
Deep retrofit of house	0.6
Switch to hybrid car	0.5
Plant-based diet	0.4
Recycle	0.15
Avoid a European flight per year	0.08
Hang-dry clothes	0.06
Buy LEDs	0.01
Reuse plastic bag 1,000 times	0.01

Source: Founders Pledge, 2020

In 2017, Lund University produced similar, though not identical, results. Want to cut your greenhouse gas emissions? Have a smaller family than you would otherwise have done, don't own a car, don't fly, and go vegan. But the researchers, Seth Wynes and Kimberly Nicholas, also examined Canadian high school science textbooks, and government resources on climate change from the EU, the US, Canada and Australia. These focused on low- or moderate-impact actions, like recycling. The guidance didn't match the facts.

To be fair, it's not as simple as the figures suggest. Some things are a lot harder for individuals to do than others, often for complex reasons. Take having fewer kids than you would have liked. (Biological children, that is: adoption doesn't add anyone to the planet.) For some people, that's a fundamentally different choice from quitting driving or flying.

Of course, this doesn't apply to everyone. That, too, is important to stress. Even in the twenty-first century, in societies where women can be politicians and sports stars, leaders of movements and corporations, we are still widely judged on whether or not we have 'achieved' motherhood. Want to improve women's lives *and* contribute to climate justice? Then challenge these norms. Nor am I saying that individuals and couples in rich, polluting states shouldn't consider the carbon tab when deciding how many kids to have. Many should. Plenty of us already do.

But, for some, 'stopping at' one or two children would be a serious sacrifice, made so by the violations of the past. Jade Sasser, a gender studies professor, points out that women of colour might value larger families *because* of the reproductive violations heaped on their ancestors through slavery, eugenics and forced sterilization.

Moreover, these figures aren't set in stone. They depend on what happens collectively. According to Founders Pledge, the 'carbon tab' of an additional child in the US is seven tonnes a year, compared to 1.4 tonnes in France. That's because of different national policies and priorities: a child born in France, and any child *they* have, will have a lower carbon footprint than one in the US. Table 2 takes account of actual policies in comparatively climate-progressive countries, but further policies could result from concerted activism. Once again, cutting your carbon footprint matters, but policy change matters more.

HOW TO DECIDE: THE BIGGER PICTURE

For years, I've been working on this vexed question: what should each of us do about climate change? My conclusion? Focus on *what you can achieve working with others who are also motivated by climate justice.* Not on what you can do narrowly as an individual; not on what you should do if all citizens in this generation acted as *they* should. Because, as we saw in the last chapter, they won't.

That doesn't mean giving up on what, in Chapter Five, we called 'non-compliers'. It means cooperating with other 'compliers' as best you can, to try to bring about the kind of global change where everyone does their share – or, failing that, the 'next best' outcome. Our aim? The fairest we can get. The baseline? Not killing people; protecting basic interests.

But how, in practice, to do this?

Progress requires action by many, many different groups, employing different strategies, formal and informal, local to global. That includes civil society, from families to international NGOs. It includes conservation volunteers and lifestyle movements, entrepreneurs, investors in renewables, and many more. So the first thing for you or me to ask is, *what works*? Which groups make a difference? How do the different efforts complement (or undermine) each other? Which of them are cutting global emissions, aiding adaptation, *and* challenging the patterns of oppression and misrecognition that have got us here? Is there anything that should be happening, and isn't?

The second thing to ask is where you belong, in this medley of cooperation and coordination. What are your skills? Are you a scientist, a technician, a lawyer, an orator, a natural leader? Where do you already have a position or influence? And how difficult are these things for you? How much would you have to give up? (The easier it is for you to do something, the more of it you can do before it becomes *too* hard.)

'Everything for Notre Dame, nothing for Les Misérables', read a sign at a 2019 *gilet jaune* protest in Paris. More than a billion dollars had been donated to repair the burnt cathedral, while French workers, the protestors said, struggled to feed their families. Meanwhile, Cyclone Idai had killed hundreds in Mozambique, devastated homes, and stripped the country of food. Aid agencies were struggling to raise funds.

Susan Moeller, professor of media and international affairs, compared reactions to the 2004 Boxing Day Tsunami in South-East Asia with those to other tragedies. Media coverage, she says, drives donations. In this case, the coverage was enormous, partly because so many Western tourists were affected. Vivid images were immediately available; stories of 'their own' personalized the tragedy for US and European viewers.

With non-humans, cuteness, rather than actual need, seems to be the deciding factor. According to the ecologist Ernest Small, animals perceived as attractive, entertaining or useful to us receive the most conservation dollars.

This affects what you or I should do. If everyone motivated to help were to ask the questions above, and answer them accurately, this discussion could end there. But they won't, so you should adjust for that. Ask which key areas are likely to be neglected, by those with money and expertise, and consider supporting them. This might mean doing something less central to your own skillset, or harder for you, because others aren't doing it at all. We must also be flexible, in our non-ideal world. Progress – social, political, technological – will emerge in some areas, and not in others. Be prepared to change course.

Here's a concern. If you waited to find out every detail above, you'd never do anything. It's a valid point. In practice, this has to be about approximating. But you know your own skills; you know what is harder or easier for you. Research has been done already to make the first step easier. NGOs, effective activism organizations and academics have examined what works and what doesn't. There are also some fundamental questions you should *always* ask. Do the groups you might support care about basic rights? Do they recognize the needs of different groups: races, genders, generations, even species? Do they listen to those who are most at risk, who have so frequently been ignored?

INCLUSIVE ACTION

'Too often our experiences are overlooked by an environment movement that is predominantly white and middle class,' Anjali Raman-Middleton, a 17-year-old clean air campaigner, told the *Guardian*. Her schoolmate, Ella Kissi-Debrah, another child of colour, died from air pollution in 2013, aged nine.

It's hardly surprising that Black communities distrust white environmentalists, if those environmentalists appear to care more for an idealized 'wild space', or for future generations of 'people like them', than for the living, choking people of colour who are already the victims of environmental harms. To be just, climate campaigning must be inclusive. It must make oppressed voices heard. Privileged activists must listen rather than tell, learn as well as do.

This applies to *how* we protest, as well as *what* the protests demand. Take activism by risking arrest. It's a valid strategy, says the grassroots collective Wretched of the Earth, but only if you're cushioned by advantage. Many victims of climate injustice have been exposed for generations to arbitrary arrest and violence. White activists must use their privilege to highlight that, too, and must embrace diverse tactics.

WHEN TO STOP?

'I got so burned out and fed up with everything I wanted to crawl into a hole and never do anything again.'

As youth activist Jamie Margolin realizes, all this is hard to do. There are benefits, even in the short term. As the IPCC points out, active travel and plant-based eating come with health gains, via better diet and fitness. Activism can be good for mental health. Playing the ostrich – carrying on, eyes closed and beefburger in hand – isn't helpful, even for the person doing it. Working with others, being part of a community, changing how you live and doing *something* positive, can ease the devastating anxiety caused by climate change.

But there are costs too.

In the year 2019, 212 environmental and land activists were murdered. According to Global Witness, which compiled the figures, the majority were in just two countries: Colombia and the Philippines. For most activists in the global north, the stakes are lower.

But they can be high enough. Arrests, legal action, social media hate campaigns. Then there are more mundane costs. Campaigning takes time, money, physical and mental effort. It takes an emotional toll. Each of us has other ties and responsibilities too: jobs and community projects, families and friends.

Morality, which is about protecting and respecting central interests, must leave space for its agents to live their own lives. Most of us agree on that. So perhaps this is another reason for inaction – for making this 'not my problem' – to add to those in Chapter One. 'It's just too hard.' But that, too, can only go so far.

What *should* each of us sacrifice? As philosophers put it, how demanding are our climate justice duties? That's an open question.

Morality is strict when it comes to causing serious harm, directly and individually. You mustn't kill or maim others. You mustn't take their homes or their children, or make them sick. You mustn't do these things even if it's really hard for you not to. But this rule is also fairly straightforward to obey. Just *don't do* those things.

Climate harms are different. They're what the philosopher Judith Lichtenberg calls 'new harms': those that most of us help to make worse, all the time. Every coffee, every burger, every time we get in a car or turn on the heating. What we should do in response is also complex. It's not as simple as 'not doing' all those individual actions, even if we could. It could require endless work.

And, to state the obvious, *we* are different. We all have different incomes, different abilities, more or less time. Some of us have caring responsibilities, often all-consuming. Others live with disability or illness, or are the victims of discrimination piled on them from birth. Some live relatively sheltered (so far) from the impacts of climate change; others have already felt the flames.

So where do we stop?

At the very least, each of us should make minor sacrifices to fight climate injustice. How much effort does it take to sign an online petition, or send an email to your local politician? Who really *needs* another expensive coffee in its throwaway cup, when the money could go to a climate justice charity? Even setting aside our complicity in collective *harm* – even if this was only a matter of fulfilling our basic human duty to save the lives we can easily save – we should do this much. It's a bare minimum: a truth (almost) universally acknowledged, albeit inadequately fulfilled.

But this bare minimum isn't enough. It wouldn't be enough even if we weren't complicit in a greater, collective harm. As we've seen already, we should do more even if others do nothing, because the lives at stake are more important than that kind of fairness. And, of course, most of us *are* complicit. These are combined harms, structural harms. That demands more than trivial inconvenience.

So how about this? We should each keep trying until the costs to us are 'significant'. Of course, we don't know exactly what this useful term means in practice, though we probably have an intuitive idea. Perhaps a significant cost is anything that interferes with a central human interest, even temporarily. Breaking my leg running to a climate march, spending months away from those I love. Perhaps we should draw the line at anything that severely disrupts our life projects, even if they're not permanently undermined.

This gets complicated. If we chose our life projects, it's not so clear that we can appeal to them as a justification for not fulfilling moral duties. Is it too demanding to turn down your dream job because it involves flying from London to Hong Kong every month? Or to refuse to work the long hours that could mean a more rewarding role? Not necessarily, according to political theorist Chiara Cordelli. If your career eats up all your hours and means endless plane trips, leaving you unable to contribute to climate justice, perhaps you should have picked another path in life.

But the argument comes with caveats. Certainly, from a climate justice perspective, we can criticize anyone who freely chooses a career in the oil industry. But an awful lot of jobs involve long hours. School-leavers or graduates, struggling in an inclement job market, would scoff at the idea that they have a smorgasbord of rewarding occupations, laid out for the choosing. Even in better times, our life choices are rarely completely free. Circumstances push us. Social norms squeeze options. That's why, ultimately, climate justice is about changing the circumstances, challenging the norms.

So 'significant' is hard to define. But is it enough? That, too, is up for debate. Climate change is a global, intergenerational emergency, an existential threat. Perhaps we should think of it as such, contemplating sacrifices closer to those demanded of individuals in wartime, or in the global pandemic. Brave activists have already given their lives to right this collective wrong. If this is beyond the call of justice – is what philosophers call supererogatory – should we throw everything *else* we have, everything but our own survival, at this disaster?

The truth is, I don't know. But I can say two things. One is this. The fight for climate justice is an ongoing campaign, not a one-off battle. Self-preservation matters. Self-care matters. It matters for you, and it matters for the world you want to create. You cannot become a better activist by destroying yourself.

The other is this, at the other extreme. Burnout is a danger for those, like Jamie Margolin, who have committed themselves, heart and soul, to the pursuit of justice. But they are a minority. Most people are a long way in the other direction, not even making minor sacrifices. Most of us living comfortable lives in affluent countries don't need to know *exactly* how much we should do. Because we already know this: it's more than we do now.

Conclusion: Key points

It's November 2021. I write this in Edinburgh, the city I love and live in. On my computer screen is a map produced by Climate Central, a non-profit news organization, marking the areas expected to vanish under rising seas. I push the dial to 2050 and Edinburgh's coast is rimmed red. Great splotches over much of Leith, almost all of Musselburgh. So many homes, so many plans and hopes, lost to our changing climate.

I zoom out. Edinburgh gets smaller, the red marks almost too tiny to see. I study a bright patch on the east coast of England; a felt-tip line along the shore from northern France to high into Denmark. For each community, a tragedy. And yet these are rich countries, which have long gained from fossil fuels and can pay for adaptation. Bangladesh has not, and cannot, but huge swathes of its land are marked by this invidious threat. Lives and livelihoods dragged under water. Kiribati has not profited either, and it is almost all red, on my map.

I'd like to look away. From reality. From this bleak glimpse of the next few decades. But that's not an option now, for any of us. So I do the opposite. I send this book out into the world, full of hard truths and arguments which, however unpalatable, are grounded in basic morality. A book not simply about climate change, but about climate *justice*. What it looks like – how radically different from the world we're in now – and why it should matter, to each of us.

I hope it will make you not only think, but act. I hope you will use and adapt these arguments to persuade others to act. And if you take away only a few key highlights from these pages, please take these:

1. This should not be controversial. It's become politically polarized, but climate justice is about the bare minimum we owe to one another. It's about people being able to lead decent lives, not starving or drowning or dying of malaria, now or in the future. It's about not killing our fellow human beings.
2. Climate change does not destroy at random. Climate injustice is racial injustice, gender injustice. Those with most at stake, who are least responsible for climate harms, are losing everything, and they are losing it

because of colonialism, slavery, oppression, and systematic disregard for basic human rights.

3 Climate justice is multispecies justice. Climate change destroys biodiversity. It drives species to extinction and kills individual animals. This is an injustice, in its own right. It's also inseparable from what we are doing to each other, as human beings.

4 Climate justice means systematic change. It requires participation: global, intersectional, and intergenerational. It requires mitigation, adaptation and compensation. Polluters must pay for this, unless they're too poor. The rich must pay, especially if they're rich on the back of past injustice. The most vulnerable must not be made to pay.

5 In practice, this isn't happening. Major polluters refuse to do their bit: governments, corporations and individuals. The best option? Persuade or require them to comply. Until then, everyone must choose between lesser evils. Other wealthy players should (but probably won't) pick up the slack.

6 Individuals feel powerless but have collective power. We should work with motivated others to challenge climate injustice, through inclusive activism, low-carbon movements, renewable and adaptation industries. In a world in crisis, each of us can – and must – be a responsible global citizen.

FURTHER READING

FOR THE GENERAL READER

Berners-Lee, M. (2019), *There Is No Planet B: A Handbook for the Make or Break Years*. Cambridge: Cambridge University Press.

Celermajer, D. (2021), *Summertime: Reflections on a Vanishing Future*. London: Hamish Hamilton.

Johnson, A., and K. Wilkinson (2020), *All We Can Save: Truth, Courage, and Solutions for the Climate Crisis*. New York: One World.

Kalmus, P. (2017), *Being the Change: Live Well and Spark a Climate Revolution*. Gabriola Island: New Society Publishers.

Klein, N. (2014), *This Changes Everything: Capitalism versus the Planet*. New York: Penguin.

Mann, M. (2021), *The New Climate War: The Fight to Take Back our Planet*. Melbourne and London: Scribe.

Maslin, M. (2021), *How to Save Our Planet: The Facts*. London: Penguin.

Nakate, V. (2021), *A Bigger Picture: My Fight to Bring a New African Voice to the Climate Crisis*. London: One Boat.

Oreskes, N., and E. Conway (2010), *Merchants of Doubt: How a Handful of Scientists Obscured the Truth on Issues from Tobacco Smoke to Global Warming*. New York: Bloomsbury.

Shue, H. (2021), *The Pivotal Generation: Why We Have a Moral Responsibility to Slow Climate Change Right Now*. Princeton: Princeton University Press.

FOR STUDENTS

Cripps, E. (2013), *Climate Change and the Moral Agent: Individual Duties in an Interdependent World*. Oxford: Oxford University Press.

Gardiner, S. (2011), *A Perfect Moral Storm: The Ethical Tragedy of Climate Change*. Oxford and New York: Oxford University Press.

Gardiner, S., S. Caney, D. Jamieson and H. Shue (eds.) (2010), *Climate Ethics: Essential Readings*. Oxford and New York: Oxford University Press.

Godfrey, P., and D. Torres (eds.) (2016), *Systematic Crises of Global Climate Change: Intersections of Race, Class and Gender*. Oxford and New York: Routledge.

Heyward, C., and D. Roser (2016), *Climate Justice in a Non-Ideal World*. Oxford: Oxford University Press.

Schlosberg, D. (2007), *Defining Environmental Justice: Theories, Movements and Nature*. New York: Oxford University Press.

Shue, H. (2014), *Climate Justice: Vulnerability and Protection*. Oxford: Oxford University Press.

BIBLIOGRAPHY

Advertising Standards Agency (2020), 'ASA Ruling on Ryanair Ltd t/a Ryanair Ltd'. London.

Al Jazeera (2021, 19 April), 'US Miners Union Backs Shift from Coal for Renewable Energy Jobs'. Doha: Al Jazeera Media Network.

Almond, R., M. Grooten and T. Petersen (2020), 'Living Planet Report 2020: Bending the Curve of Biodiversity Loss'. Gland: WWF.

Amnesty International, Environmental Rights Action, and Friends of the Earth (2020), 'No Clean-Up, No Justice: An Evaluation of the Implementation of UNEP's Environmental Assessment of Ogoniland, Nine Years On'. London, Brussels, Amsterdam and Benin City.

Atherton, R. (2020, 3 March), 'Climate Anxiety: Survey for Newsround Shows Children Losing Sleep over Climate Change and the Environment', *Newsround*. London: BBC.

Atteridge, A., and C. Strambo (2020), 'Seven Principles to Realize a Just Transition to a Low-Carbon Economy'. Stockholm: Stockholm Environment Institute.

Bearak, M. (2020, 7 August), '"We hope that our land will become ours again"', the *Independent*. London: Independent News and Media.

Becker, L. (1986), *Reciprocity*. London and New York: Routledge & Kegan Paul.

Benoit, P. (2020), 'A Luxury Carbon Tax to Address Climate Change and Inequality: Not All Carbon Is Created Equal',

Ethics and International Affairs. New York: Cambridge University Press.

Bentham, J. (1789), *An Introduction to the Principles of Morals and Legislation* (1876 ed.). Oxford: Clarendon Press.

Brady, A., A. Torres and P. Brown (2019, 9 April), 'What the Queer Community Brings to the Fight for Climate Justice', *Grist*. Seattle: Grist Magazine.

Broome, J. (2012), *Climate Matters: Ethics in a Warming World*. New York: W. W. Norton.

Brulle, R., M. Aronczyk and J. Carmichael (2020), 'Corporate Promotion and Climate Change: An Analysis of Key Variables Affecting Advertising Spending by Major Oil Corporations, 1986–2015', *Climatic Change*, 159 (1): 87–101.

Bugden, D. (2020), 'Does Climate Protest Work? Partisanship, Protest, and Sentiment Pools', *Socius*, 6: 1–13.

Bush, G. (2001, 11 June), 'President Bush Discusses Global Climate Change', *White House Press Archives*. Washington, D.C.

Butler-Sloss, S., K. Bond and H. Benham (2021), *Spiralling Disruption: The Feedback Loops of the Energy Transition*. London: Carbon Tracker Initiative.

Caney, S. (2008), 'Human Rights, Climate Change, and Discounting', *Environmental Politics*, 17 (4): 536–55.

Caney, S. (2009), 'Climate Change, Human Rights, and Moral Thresholds', in S. Humphreys (ed.), *Human Rights and Climate Change*. Cambridge: Cambridge University Press.

Caney, S. (2010), 'Climate Change and the Duties of the Advantaged', *Critical Review of International Social and Political Philosophy*, 13 (1): 203–28.

Caney, S. (2016), 'Six Ways of Responding to Non-Compliance', in C. Heyward and D. Roser (eds.), *Climate Change and Non-Ideal Theory*. Oxford: Oxford University Press.

Caney, S. (2016), 'The Struggle for Climate Justice in a Non-Ideal World', *Midwest Studies in Philosophy*, 40 (1): 9–26.

Carrington, D. (2021, 6 May), 'How to Spot the Difference between a Real Climate Policy and a Greenwashing Guff', the *Guardian*. London: Guardian Media Group.

Ceballos, G., P. Ehrlich and P. Raven (2020), 'Vertebrates on the Brink as Indicators of Biological Annihilation and the Sixth Mass Extinction', *Proceedings of the National Academy of Sciences*, 117 (24): 13596–602.

Celermajer, D. (2021), *Summertime: Reflections on a Vanishing Future*. London: Hamish Hamilton.

Chenoweth, E., and M. Stephan (2012), *Why Civil Resistance Works: The Strategic Logic of Nonviolent Conflict*. New York: Columbia University Press.

Citizens' Assembly (accessed 12 September 2021), 'Recommendations on how the State can make Ireland a leader in tackling climate change', https://2016–2018.citizensassembly.ie/en/How-the-State-can-make-Ireland-a-leader-in-tackling-climate-change/Recommendations/

Clayton, S., C. M. Manning, K. Krygsman and M. Speiser (2017), *Mental Health and Our Changing Climate: Impacts, Implications, and Guidance*. Washington, D.C.: American Psychological Association and ecoAmerica.

Climate Accountability Institute (2019), 'Carbon Majors: Update of Top Twenty Companies 1965–2017'. Snowmass, Co.

Climate Action Tracker (2021), 'Global Update: Climate target updates slow as science demands action'. Berlin: Climate Analytics and New Climate Institute.

Climate Action Tracker (2021), 'Glasgow's 2030 credibility gap: net zero's lip service to climate action'. Berlin: Climate Analytics and New Climate Institute.

Climate Central (accessed 26 May 2021), 'Coastal Risk Screening Tool: Land Projected to be Below Annual Flood Level in 2050', https://coastal.climatecentral.org

Climate Coalition, Priestley International Centre for Climate, and UK Health Alliance on Climate Change (2021), 'This Report Comes with a Health Warning: The Impacts of Climate Change on Public Health'.

Climate Justice Alliance (accessed 7 July 2021), 'Just Transition: A Framework for Change', https://climatejusticealliance.org/just-transition

ClimateWatch (accessed 11 July 2021), 'Historical GHG Emissions', https://www.climatewatchdata.org/ghg-emissions

Coffey, C., P. Espinoza Revollo, R. Harvey, M. Lawson, A. Parvez Butt, K. Piaget, D. Sarosi and J. Thekkudan (2020), 'Time to Care: Unpaid and Underpaid Care Work and the Global Inequality Crisis'. Oxford: Oxfam International.

Committee on Climate Change (2019), 'Net Zero: The UK's Contribution to Stopping Global Warming'. London.

Cordelli, C. (2018), 'Prospective Duties and the Demands of Beneficence', *Ethics*, 128 (2): 373–401.

Crenshaw, K. (1989), 'Demarginalizing the Intersection of Race and Sex: A Black Feminist Critique of Antidiscrimination Doctrine, Feminist Theory and

Antiracist Politics', *University of Chicago Legal Forum*, 1: 139–67.

Cullity, G. (1995), 'Moral Free Riding', *Philosophy and Public Affairs*, 24 (1): 3–34.

Di Chiro, G. (2017), 'Welcome to the White (M)Anthropocene?', in S. MacGregor (ed.), *Routledge Handbook of Gender and Environment*. Oxford: Routledge.

Dickens, C. (1843), *A Christmas Carol*. London: Chapman & Hall.

Dimitrov, R. S. (2016), 'The Paris Agreement on Climate Change: Behind Closed Doors', *Global Environmental Politics*, 16 (3): 1–11.

Doran, P., and M. Kendall Zimmerman (2009), 'Examining the Scientific Consensus on Climate Change', *Eos, Transactions American Geophysical Union*, 90: 22–3.

DuVernay, A. (2016), 'Race, Social Class, and Disasters: The Katrina Version of Reality', in P. Godfrey and D. Torres (eds.), *Systematic Crisis of Global Climate Change: Intersections of Race, Class and Gender*. London and New York: Routledge, 285–95.

Effective Activist (accessed 26 May 2021), *Components of Effective Activism*, https://effectiveactivist.com/activism-components

Ehrlich, P. R., and J. P. Holdren (1972), 'A Bulletin Dialogue on "The Closing Circle": Critique', *Bulletin of the Atomic Scientists*, 28 (5): 16–27.

Environmental Justice Network (1991), 'The Principles of Environmental Justice'. Washington, D.C.

Environmental Protection Agency (2014), 'National Air Toxics Assessment: 2014 NATA Map'. Washington, D.C.

Evelyn, K. (2020, 29 January), '"Like I wasn't there": Climate Activist Vanessa Nakate on Being Erased from a Movement', the *Guardian*. London: Guardian Media Group.

Fisher, D. R. (2019), 'The Broader Importance of #FridaysForFuture', *Nature Climate Change*, 9 (6): 430–1.

Fisher, D., and S. Nasrin (2020), 'Climate Activism and its Effects', *Wiley Interdisciplinary Reviews: Climate Change*, 12.

Fitzgerald, L. M., P. Tobin, C. Burns and P. Eckersley (2021), 'The "Stifling" of New Climate Politics in Ireland', *Politics and Governance*, 9 (2): 41–50.

Fraser, J., V. Pantesco, K. Plemons, R. Gupta and S. J. Rank (2013), 'Sustaining the Conservationist', *Ecopsychology*, 5 (2): 70–9.

Fraser, N. (1996), *Social Justice in the Age of Identity Politics: Redistribution, Recognition, and Participation*. The Tanner Lectures on Human Values. Stanford: Stanford University.

Fridays for Future (accessed 25 May 2021), 'September 25 – Global Day of Climate Action', https://fridaysforfuture.org/september25

Gabour, J. (2015, 27 August), 'A Katrina Survivor's Tale: "They forgot us and that's when things started to get bad"', the *Guardian*. London: Guardian Media Group.

Gardiner, S. (2011), *A Perfect Moral Storm: The Ethical Tragedy of Climate Change*. Oxford and New York: Oxford University Press.

Gardiner, S. (2011), 'Is No-One Responsible for Global Environmental Tragedy? Climate Change as a Challenge to Our Ethical Concepts', in D. G. Arnold (ed.), *The*

Ethics of Global Climate Change. Cambridge: Cambridge University Press, 38–59.

Gardiner, S. (2013), 'Geoengineering and Moral Schizophrenia', in W. C. G. Burns and A. L. Strauss (eds.), *Climate Change Geoengineering: Philosophical Perspectives, Legal Issues, and Governance Frameworks*. New York: Cambridge University Press.

Gardiner, S. (2014), 'A Call for a Global Constitutional Convention Focused on Future Generations', *Ethics & International Affairs*, 28 (3): 299–315.

Giuffrida, A. (2021, 5 August), 'Eight Dead as Wildfires Continue to Rage across Southern Europe', the *Guardian*. London: Guardian Media Group.

Global Assembly (accessed 16 September 2021), 'About', https://globalassembly.org/about

Global Witness (2020), 'Defending Tomorrow: The Climate Crisis and Threats against Land and Environmental Defenders'. London.

Godfrey, P., and Torres, D. (2016), *Systematic Crises of Global Climate Change: Intersections of Race, Class and Gender*. Oxford and New York: Routledge.

Goldberg, M. H., A. Gustafson, S. A. Rosenthal and A. Leiserowitz (2021), 'Shifting Republican Views on Climate Change through Targeted Advertising', *Nature Climate Change*.

Gosseries, A. (2004), 'Historical Emissions and Free-Riding', *Ethical Perspectives*, 11 (1): 36–60.

Gössling, S., and A. Humpe (2020), 'The Global Scale, Distribution and Growth of Aviation: Implications for Climate Change', *Global Environmental Change*, 65: 102–94.

Green Climate Fund (accessed 26 May 2021), 'Resource Mobilization', https://www.greenclimate.fund/about/resource-mobilization

Halliday, E. (2020, 1 October), 'Forced Abortions and Secret Sterilization: How China has Abused Uighar Women for Decades', *New Statesman*.

Halstead, J., and J. Ackva (2020), *Climate & Lifestyle Report*. London: Founders Pledge.

Hare, C. (2007), 'Voices from Another World: Must We Respect the Interests of People Who Do Not, and Will Never, Exist?', *Ethics*, 117 (3): 498–523.

Harris, D. (2016), 'The Political Ecology of *Pachamama*', in P. Godfrey and D. Torres (eds.), *Systematic Crises of Global Climate Change: Intersections of Race, Class and Gender*. Oxford and New York: Routledge, 186–98.

Harris, P. (2007, 4 March), 'Relatives Demand Justice as Police Go on Trial over Katrina Killings', the *Guardian*. London: Guardian Media Group.

Harvey, F. (2019, 2 December), 'Greta Thunberg Says School Strikes Have Achieved Nothing', the *Guardian*. London: Guardian Media Group.

Heartland Institute, The (accessed 2 December 2021), 'Why Scientists Disagree About Global Warming', https://www.heartland.org/publications-resources/publications/why-scientists-disagree-about-global-warming

Heglar, M. (2019, 18 February), 'Climate Change Isn't the First Existential Threat', *Medium*. San Francisco: A Medium Corporation.

Heglar, M. (2020, 16 June), 'We Don't Have to Halt Climate Action to Fight Racism', *Huffington Post*. New York: Verizon.

Heglar, M. (2020, 15 November), 'Get in Losers, We're Going GreenTrolling', *Hot Take Newsletter*. HotTake.

Helliwell, J., R. Layard, J. Saches and J.-E. Neve (eds.), *World Happiness Report*. Paris and New York: Sustainable Development Solutions Network.

Henley, J. (2021), 'Turkey Flood Deaths Rise as Fresh Fires Erupt on Greek Island of Evia', the *Guardian*. London: Guardian Media Group.

Hettinger, N. (2018), 'Naturalness, Wild-Animal Suffering, and Palmer on Laissez-Faire', *The Ethics Forum*, 13 (1): 65–84.

Hickman, C. (2020), 'We Need to (Find a Way to) Talk About … Eco-Anxiety', *Journal of Social Work Practice*, 34 (4): 411–24.

Hickman, C., E. Marks, P. Pihkala, S. Clayton, E. Lewandowski, E. Mayall, B. Wray, C. Mellor and L. van Susteren (2021), 'Young People's Voices on Climate Anxiety, Government Betrayal and Moral Injury: A Global Phenomenon', *The Lancet* (preprints).

Hobbes, T. (1651), *Leviathan* (1996 ed.), Cambridge: Cambridge University Press.

Hoffman, J. S., V. Shandas and N. Pendleton (2020), 'The Effects of Historical Housing Policies on Resident Exposure to Intra-Urban Heat: A Study of 108 US Urban Areas', *Climate*, 8 (1): 12.

Holland, B. (2012), 'Environment as a Meta-Capability: Why a Dignified Human Life Requires a Stable Climate System', in A. Thompson and J. Bendik-Keymer (eds.), *Ethical Adaptation to Climate Change: Human Virtues of the Future*. Cambridge, MA, and London: MIT Press, 145–64.

Idso, C., R. Carter and S. Singer (2015), 'Why Scientists Disagree About Global Warming: The NIPCC Report on Scientific Consensus'. Arlington: The Heartland Institute.

Ingram, S. (2019), 'A Gathering Storm: Climate Change Clouds the Future of Children in Bangladesh'. Dhaka: UNICEF Bangladesh.

International Energy Agency (2021), 'Net Zero by 2050'. Paris.

International Energy Agency (accessed 26 May 2021), 'Energy Subsidies: Tracking the Impact of Fossil Fuel Subsidies'. Paris.

IPCC (1992), 'Climate Change: The IPCC 1990 and 1992 Assessments'. Geneva: Intergovernmental Panel on Climate Change.

IPCC (2014), 'Climate Change 2014: Synthesis Report. Contribution of Working Groups I, II and III to the Fifth Assessment Report of the Intergovernmental Panel on Climate Change'. Geneva: IPCC.

IPCC (2014), 'Climate Change 2014: Impacts, Adaptation, and Vulnerability, Part A: Global and Sectoral Aspects. Contribution of Working Group II to the Fifth Assessment Report of the Intergovernmental Panel on Climate Change'. Cambridge and New York: Cambridge University Press.

IPCC (2018), 'Global Warming of 1.5°C: An IPCC special report on the impacts of global warming of 1.5°C above pre-industrial levels and related global greenhouse gas emissions pathways, in the context of strengthening the global response to the threat of climate change, sustainable development, and efforts to eradicate poverty'. Cambridge and New York: Cambridge University Press.

IPCC (2021), 'Climate Change 2021: The Physical Science Basis. Contribution of Working Group I to the Sixth

Assessment Report of the Intergovernmental Panel on Climate Change'. Cambridge and New York: Cambridge University Press.

Kalhoefer, K. (2016), 'Study: CNN Viewers See Far More Fossil Fuel Advertising Than Climate Change Reporting'. Washington, D.C.: Media Matters for America.

Kant, I. (1785), *Groundwork for the Metaphysics of Morals* (2019 ed.). Oxford: Oxford University Press.

Kartha, S., E. Kemp-Benedict, E. Ghosh, A. Nazareth and T. Gore (2020), 'The Carbon Inequality Era: An Assessment of the Global Distribution of Consumption Emissions among Individuals from 1990 to 2015 and Beyond'. Stockholm and Oxford: Stockholm Environment Institute and Oxfam International.

Klein, N. (2014), *This Changes Everything: Capitalism versus the Planet*. New York: Penguin.

Kosanic, A., J. Petzold, A. Dunham and M. Razanajatovo (2019), 'Climate Concerns and the Disabled Community', *Science*, 366: 6466, 698–9.

Kumari Rigaud, K., A. de Sherbinin, B. Jones, J. Bergmann, V. Clement, K. Ober, J. Schewe, S. Adamo, B. McCusker, S. Heuser and A. Midgley (2018), 'Groundswell: Preparing for Internal Climate Migration'. Washington, D.C.: The World Bank.

Kutz, C. (2000), *Complicity: Ethics and Law for a Collective Age*. Cambridge and New York: Cambridge University Press.

Lakner, C., N. Yonzan, D. Mahler, R. Aguiler and H. Wu (2021, 11 January), 'Updated Estimates of the Impact of COVID-19 on Global Poverty: Looking Back at 2020 and the Outlook for 2021'. Washington, D.C.: World Bank.

Lavelle, M. (2018, 29 October), 'Big Oil Has Spent Millions of Dollars to Stop a Carbon Fee in Washington State', *Inside Climate News*. New York.

Lawrence, P. (2019), 'Representation of Future Generations', in A. Kalfagianni, D. Fuchs and A. Hayden (eds.), *Routledge Handbook of Global Sustainability Governance*. London: Routledge, 88–99.

Lazarus, O., S. McDermid and J. Jacquet (2021), 'The Climate Responsibilities of Industrial Meat and Dairy Producers', *Climatic Change*, 165 (1): 30.

Lewis, S. (2021, 3 March), 'The Climate Crisis Can't Be Solved by Carbon Accounting Tricks', the *Guardian*. London: Guardian Media Group.

Lexington Herald Leader (2012, 18 March), 'Joel Pett: The Cartoon Seen "round the world"', Lexington: McClatchy Company.

Lichtenberg, J. (2010), 'Negative Duties, Positive Duties, and the "New Harms"', *Ethics*, 120 (3): 557–78.

Long, J., N. Harré and Q. Atkinson (2014), 'Understanding Change in Recycling and Littering Behavior Across a School Social Network', *American Journal of Community Psychology*, 53 (3–4): 462–74.

MacGregor, S. (2014), 'Only Resist: Feminist Ecological Citizenship and the Post-politics of Climate Change', *Hypatia*, 29 (3): 617–33.

Magnani, A. (2021, 19 April), 'Andean Glaciers Are Melting, Reshaping Centuries-Old Indigenous Rituals', *National Geographic*. Washington, D.C.

Mavisakalyan, A., and Y. Tarverdi (2019), 'Gender and Climate Change: Do Female Parliamentarians Make a Difference?', *European Journal of Political Economy*, 56: 151–64.

May, L. (1992), *Sharing Responsibility*. Chicago and London: University of Chicago Press.

McCright, A., and R. Dunlap (2011), 'Cool Dudes: The Denial of Climate Change among Conservative White Males in the United States', *Global Environmental Change*, 21 (4): 1163–72.

McGreal, C. (2012, 4 April), 'Five New Orleans Police Officers Sentenced in Hurricane Katrina Killings', the *Guardian*. London: Guardian Media Group.

McMahan, J. (2008), 'Challenges to Human Equality', *Journal of Ethics*, 12: 81–104.

Mellor, D. J. (2016), 'Updating Animal Welfare Thinking: Moving beyond the "Five Freedoms" towards "A Life Worth Living"', *Animals: An Open Access Journal from MDPI* 6 (3).

Mill, J. S. (1859), *On Liberty*. London: John W. Parker & Son.

Mills, C. (2005), '"Ideal Theory" as Ideology', *Hypatia*, 20 (3): 165–84.

Moeller, S. (2006), '"Regarding the Pain of Others": Media, Bias and the Coverage of International Disasters', *Journal of International Affairs*, 59 (2): 173–96.

Molloy, M. (2013, 5 December), 'Video: Dramatic Rescue of Elderly Driver Stuck in Sinking Car Captured on Camera', *Metro*.

Monir, F. (2019, 17 December), 'Circling Cyclones, Tigers and Salt Water: Fears of a Community on the Frontline of the Climate Crisis', Bangladesh: Oxfam blog.

Monroe, I. (2015, 2 September; updated 2 September 2016), 'Hurricane Katrina's Struggling Black Gay Community', *Huffington Post*. New York: Verizon.

Myers, N. (2002), 'Environmental Refugees: A Growing Phenomenon of the 21st Century', *Philosophical Transactions of the Royal Society of London. Series B, Biological Sciences*, 357 (1420): 609–13.

Nakate, V. (2021, 8 February), 'A 2C Hotter World is a Death Sentence for Countries Like Mine', the *Independent*. London: Independent News and Media.

Nefsky, J. (2017), 'How You Can Help, Without Making a Difference?', *Philosophical Studies*, 174: 2743–67.

Nellemann, C., R. Verma and L. Hislop (eds.) (2011), 'Women at the Frontline of Climate Change: Gender Risks and Hopes: A Rapid Response Assessment'. Nairobi: United National Environment Programme and GRID-Arendal.

Nozick, R. (1974), *Anarchy, State and Utopia*. Oxford: Blackwell.

Nussbaum, M. (2000), *Women and Human Development: The Capabilities Approach*. Cambridge: Cambridge University Press.

Nussbaum, M. (2006), *Frontiers of Justice: Disability, Nationality, Species Membership*. Cambridge, MA, and London: Harvard University Press.

O'Neill, D., A. Fanning, W. Lamb and J. Steinberger (2018), 'A Good Life For All Within Planetary Boundaries', *Nature Sustainability*, 1: 88–95.

O'Neill, R., and K. Hughes (2014), 'The State of England's Chalk Streams'. Woking: WWF-UK.

Oreskes, N. (2004), 'Beyond the Ivory Tower: The Scientific Consensus on Climate Change', *Science*, 306 (5702): 1686.

Oreskes, N., and E. Conway (2010), *Merchants of Doubt: How a Handful of Scientists Obscured the Truth on Issues from Tobacco Smoke to Global Warming*. New York: Bloomsbury.

Our Children's Trust (accessed 12 September 2021), 'Youth v. Gov: Julian v. US', https://www.ourchildrenstrust.org/juliana-v-us

Oxfam International (2005), 'The Tsunami's Impact on Women'. Oxford: Oxfam Briefing Note.

Oxford Martin School (accessed 26 May 2021), Oxford Geoengineering Programme, https://www.oxfordmartin.ox.ac.uk/geoengineering

Parfit, D. (1984), *Reasons and Persons* (1987 corrected reprint ed.). Oxford: Clarendon Press.

People's Daily (2011, 28 October), '400 Million Births Prevented by One-Child Policy'. Beijing: People's Daily Online.

Pepper, A. (2019), 'Adapting to Climate Change: What We Owe to Other Animals', *Journal of Applied Philosophy*, 36 (4): 592–607.

Petchesky, R. (1995), 'From Population Control to Reproductive Rights: Feminist Fault Lines', *Reproductive Health Matters*, 6: 152–61.

Pianta, S., and M. R. Sisco (2020), 'A Hot Topic in Hot Times: How Media Coverage of Climate Change is Affected by Temperature Abnormalities', *Environmental Research Letters*, 15 (11).

Plumwood, V. (1993), *Feminism and the Mastery of Nature*. London and New York: Routledge.

Rainforest Action Network, Banktrack, Indigenous Environment Network, Oil Change, Reclaim Finance, & Sierra Club (2021), *Banking on Climate Chaos: Fossil Fuel Finance Report 2021*. San Francisco.

Rajamani, L. (2016), 'The 2015 Paris Agreement: Interplay Between Hard, Soft and Non-Obligations', *Journal of Environmental Law*, 28 (2): 337–58.

Rajamani, L., and D. Bodansky (2019), 'The Paris Rulebook: Balancing International Prescriptiveness with National Discretion', *International and Comparative Law Quarterly*, 68 (4): 1023–40.

Rawls, J. (1971), *A Theory of Justice* (revised 1999 ed.). Oxford: Oxford University Press.

Raworth, K. (2012), 'A Safe and Just Space for Humanity'. Oxford: Oxfam Briefing Paper.

Rees, N., D. Anthony, C. Colon, A. Heikens, C. Klauth, J. Cosby and E. Garin (2015), 'Unless We Act Now: The Impact of Climate Change on Children'. New York: UNICEF.

Richards, J.-A., and S. Bradshaw (2017), 'Uprooted by Climate Change: Responding to the Growing Risk of Displacement'. Oxford: Oxfam International.

Ritchie, H., and M. Roser (2020, August), 'CO2 and Greenhouse Gas Emissions', *Our World in Data*. London and Oxford: Global Change Data Lab and Oxford Martin School.

Robinson, M., and C. Palmer (2018), *Climate Justice: A Man-Made Problem with a Feminist Solution*. London: Bloomsbury.

Roser, D., and S. Hohl (2011), 'Stepping in for the Polluters? Climate Justice under Partial Compliance', *Analyse and Kritik*, 33 (2): 477–500.

Royal College of Psychiatrists (2020, 20 November), 'The Climate Crisis is Taking a Toll on the Mental Health of Children and Young People'. London: RCP Online News.

Rozenberg, J., and S. Hallegatte (2018), 'Poor People on the Front Line: The Impacts of Climate Change on Poverty in 2030', in *Climate Justice: Integrating Economics and*

Philosophy by H. Shue and R. Kanbur (eds.). Oxford: Oxford University Press, 25–38.

Runge, C., and S. Benjamin (2007), 'How Biofuels Could Starve the Poor', *Foreign Affairs*, 86 (3): 41.

Sainato, M. (2020, 29 May), 'The Collapse of Coal: Pandemic Accelerates Appalachia Job Losses', the *Guardian*. London: Guardian Media Group.

Sasser, J. (2018), *On Infertile Ground: Population Control and Women's Rights in the Era of Climate Change*. New York: New York University Press.

Savaresi, A. (2012), 'The Human Rights Dimension of REDD', *Review of European Community & International Environmental Law*, 21 (2): 102–13.

Schlosberg, D. (2007), *Defining Environmental Justice: Theories, Movements and Nature*. New York: Oxford University Press.

Schlosberg, D., L. Collins and S. Niemeyer (2017), 'Adaptation Policy and Community Discourse: Risk, Vulnerability, and Just Transformation', *Environmental Politics*, 26 (3): 413–37.

Sen, A. (1999), *Development as Freedom*. Oxford: Oxford University Press.

Shafto, J. (2019, 23 September), 'Oil, Gas Sector Struggling to Attract, Elevate Women Amid Gender Diversity Gap'. New York: S&P Market Intelligence.

Shakespeare, W. (1603), *Hamlet* (2016 ed.). London and New York: Bloomsbury.

Sharkey, P. (2007), 'Survival and Death in New Orleans: An Empirical Look at the Human Impact of Katrina', *Journal of Black Studies*, 37 (4): 482–501.

Shiffrin, S. (1999), 'Wrongful Life, Procreative Responsibility, and the Significance of Harm', *Legal Theory*, 5: 117–48.

Shue, H. (1980), *Basic Rights: Subsistence, Affluence, and U.S. Foreign Policy*. Princeton: Princeton University Press.

Shue, H. (1992), 'The Unavoidability of Justice', in A. Hurrell and B. Kinsbury (eds.), *The International Politics of the Environment*. Oxford: Oxford University Press.

Shue, H. (1993), 'Subsistence Emissions and Luxury Emissions', *Law and Policy*, 15 (1): 39–60.

Shue, H. (1999), 'Bequeathing Hazards: Security Rights and Property Rights of Future Humans', in M. Dore and T. Mount (eds.), *Global Environmental Economics: Equity and the Limits to Markets*. Oxford: Blackwell.

Shue, H. (2019), 'Subsistence Protection and Mitigation Ambition: Necessities, Economic and Climatic', *British Journal of Politics and International Relations*, 21 (2): 251–62.

Shue, H. (2021), *The Pivotal Generation: Why We Have a Moral Responsibility to Slow Climate Change Right Now*. Princeton: Princeton University Press.

Singer, P. (1972), 'Famine, Affluence, and Morality', *Philosophy and Public Affairs*, 72 (1), 229–43.

Singer, P. (1976), 'All Animals Are Equal', in P. Singer (ed.), *Animal Ethics*. Oxford: Oxford University Press.

Sinnott-Armstrong, W. (2005), 'It's Not My Fault: Global Warming and Individual Moral Obligations', in W. Sinnott-Armstrong and R. Howarth (eds.), *Perspectives on Climate Change: Science, Economics, Politics, Ethics*. Oxford: Elsevier.

Small, E. (2011), 'The New Noah's Ark: Beautiful and Useful Species Only. Part 1. Biodiversity Conservation Issues and Priorities', *Biodiversity*, 12.

Sopoaga, E. (2014, 14 December), 'Statement Presented at the 20th Conference of Parties to the UN Framework Convention on Climate Change'. Lima: UNFCCC.

Standing Rock Sioux Tribe (accessed 26 May 2021), Treaties Still Matter: The Dakota Access Pipeline. https://americanindian.si.edu/nk360/plains-treaties/dapl

Steinberger, J. [@JKSteinberger] (2021, 13 June), 'This is not surprising . . .' https://twitter.com/JKSteinberger/status/1404163671157313543

Stern, N. (2006), *Review on the Economics of Climate Change*. London: HM Treasury.

Stichting Reclame Code (2020), 'Beslissing van de Reclame Code Commissie in de zaak van: E.C. Stam, wonende te Rotterdam, klager tegen: KLM Royal Dutch Airlines, devestigd te Luchthaven Schiphol, verweerder'. Amsterdam.

Stockholm Environment Institute, IISD, ODI, E3G, and UN Environment Programme (2020), 'The Production Gap: The Discrepancy between Countries' Planned Fossil Fuel Production and Global Production Levels Consistent with Limiting Warming to 1.5°C or 2°C'. Stockholm.

Supran, G., and N. Oreskes (2017), 'Assessing ExxonMobil's Climate Change Communications (1977–2014)', *Environmental Research Letters*, 12 (2).

Supran, G., and N. Oreskes (2020), 'Addendum to "Assessing ExxonMobil's Climate Change Communications (1977–2014)"', *Environmental Research Letters*, 15 (1).

Supran, G., and N. Oreskes (2021), 'Rhetoric and Frame Analysis of ExxonMobil's Climate Change Communications', *One Earth*, 4 (5): 696–719.

Survival International (accessed 24 June 2021), 'Decolonize Conservation: Indigenous People Are the Best Conservationists', https://www.survivalinternational.org/conservation

Sussman, R., M. Greeno, R. Gifford and L. Scannell (2013), 'The Effectiveness of Models and Prompts on Waste Diversion: A Field Experiment on Composting by Cafeteria Patrons', *Journal of Applied Social Psychology*, 43 (1): 24–34.

Tansey, R. (2019), 'Big Oil and Gas Buying Influence in Brussels', Brussels: Corporate Europe Observatory, Food & Water Europe, Friends of the Earth Europe, Greenpeace UK.

Taylor, M., E. Holder, D. Collyns, M. Standaert and A. Kassam (2021, 7 May), 'The Young People Taking their Countries to Court over Climate Inaction', the *Guardian*. London: Guardian Media Group.

Team, V., and E. Hassen (2016), 'Climate Change and Complexity of Gender Issues in Ethiopia', in P. Godfrey and D. Torres (eds.), *Systematic Crises of Global Climate Change: Intersections of Race, Class and Gender*. Oxford and New York: Routledge, 314–26.

Thomas, L. (2020, 24 June), 'Intersectional Environmentalism Is Our Urgent Way Forward', *Youth to the People*. Los Angeles.

Thompson, D. (2021, 21 April), 'Most Sydneysiders Still Want Government to Lead on Climate Change Adaptation'. Sydney: Western Sydney University Press Release.

Thunberg, G. (2019), *No-One is Too Small to Make a Difference*. London: Penguin.

Tschakert, P., D. Schlosberg, D. Celermajer, L. Rickards, C. Winter, M. Thaler, M. Stewart-Harawira and B. Verlie (2021), 'Multispecies Justice: Climate-Just Futures With, For and Beyond Humans', *WIRES Climate Change*, 12 (2).

United Nations (1948), 'UN Declaration of Human Rights'. Paris.

United Nations (2015), 'Paris Agreement on Climate Change', United Nations Framework Convention on Climate Change, Paris.

United Nations (2019), 'Differentiated impacts of climate change on women and men; the integration of gender considerations in climate policies, plans and actions; and progress in enhancing gender balance in national climate delegations', United Nations Framework Convention on Climate Change. Secretariat Synthesis Report.

United Nations (2020), 'Gender Composition: United Nations Framework Convention on Climate Change'. Secretariat Synthesis Report.

United Nations (accessed 9 July 2021), 'The Paris Outcome on Loss and Damage', https://unfccc.int/files/adaptation/groups_committees/loss_and_damage_executive_committee/application/pdf/ref_8_decision_xcp.21.pdf

United Nations Department of Economic and Social Affairs (accessed 25 May 2021), The 17 Goals, https://sdgs.un.org/goals

United Nations Environment Programme (2020), 'Emissions Gap Report 2020.' Nairobi.

United Nations (2021), 'Glasgow Climate Pact' (Advance unedited version), United Nations Framework Convention on Climate Change, Glasgow.

US Department of State (2019), '2019 Country Reports on Human Rights Practices: China (includes Tibet, Hong Kong, and Macau)'. Washington, D.C.

Vestal, C. (2017, 10 December), '"Katrina Brain": The Invisible Long-Term Toll of Megastorms', *Politico*. Washington, D.C.

Vidal, J. (2015, 16 December), 'How a "typo" Nearly Derailed the Paris Climate Deal', the *Guardian*. London: Guardian Media Group.

Vidal, J. (2020, 7 February), 'Armed Ecoguards Funded by WWF "beat up Congo tribespeople"', the *Guardian*. London: Guardian Media Group.

Vince, G. (2020, 12 January), 'How Scientists are Coping with "ecological grief"', the *Guardian* London: Guardian Media Group.

Vitale, A. (2019, October ed.), 'What I Learned Documenting the Last Male Northern White Rhino's Death', *National Geographic*. Washington, D.C.

Wet Tropics Management Authority (2020), 'State of Wet Tropics 2019–2020'. Cairns, Australia.

Whyte, K. P. (2014), 'Indigenous Women, Climate Change Impacts, and Collective Action', *Hypatia*, 29 (3): 599–616.

Williamson, B., F. Markham and J. Weir (2021), 'Aboriginal Peoples and the Response to the 2019–2020 Bushfires', Centre for Aboriginal Economic Policy Research, Australian National University, Canberra.

Woodward, J. (1986), 'The Non-Identity Problem', *Ethics*, 96 (4): 804–31.

World Citizens' Assembly (accessed 16 September 2021), 'Our Model', https://www.worldassembly.org/model

Wretched of the Earth (2019, 3 May), 'An Open Letter to Extinction Rebellion', *Red Pepper*. London.

WWF Australia (2020), 'Australian's 2019–2020 Bushfires: The Wildlife Toll'.

WWF-UK (2018), 'Wildlife in a Warming World: The Effects of Climate Change on Biodiversity in WWF's Priority Places'. Woking.

Wynes, S., and K. Nicholas (2017), 'The Climate Mitigation Gap: Education and Government Recommendations Miss the Most Effective Individual Actions', *Environmental Research Letters*, 12.

Xie, B., and B. Newell (2019, 18 September), 'Why Attending a Climate Strike Can Change Minds (Most Importantly Your Own)', *The Conversation*. London.

Xie, B., M. Brewer, B. Hayes, R. McDonald and B. Newell (2019), 'Predicting Climate Change Risk Perception and Willingness to Act', *Journal of Environmental Psychology*, 65: 101331.

Zero Carbon Campaign (2021), 'Carbon Tax Would Be Popular with UK Voters, Poll Suggests'. UK.

Zou, Jie Jenny (2017, 15 June), 'Oil's Pipeline to America's Schools: Inside the Fossil-Fuel Industry's Not-So-Subtle Push Into K-12 Education', Washington, D.C.: Center for Public Integrity.

ACKNOWLEDGEMENTS

I couldn't have written this book without the enthusiasm and editorial brilliance of my agent, Jaime Marshall, and my editor, Jamie Birkett. I am deeply grateful for the support of the whole team at Bloomsbury, including Robin Baird-Smith, Lizzy Ewer, Jude Drake, Sarah Jones, Rosie Parnham, Tomasz Hoskins and Richard Mason. Nor could I have done without the generosity and expertise of my colleagues, especially Elizabeth Bomberg, Claire Duncanson, Darrick Evensen, Fiona Mackay, Mihaela Mihai and Mathias Thaler. I'm very lucky to work with you. I'm lucky, too, to have been able to discuss these ideas with many superb students, from undergraduates to PhD candidates.

As a philosopher, I couldn't have chosen a more collegiate and welcoming field than climate justice and ethics. I have drawn on the ideas of many wonderful fellow scholars here. But particular thanks go to Simon Caney, Stephen Gardiner, Catriona McKinnon, David Schlosberg and Henry Shue, for inspiration and kindness. This book draws, too, on the wisdom and work of many activists and journalists. I gratefully acknowledge that.

Five accomplished academics and activists commented on a draft of this book for an 'Author Meets Critics' workshop, organized by CRITIQUE, the Centre for Ethics and Critical Thought, at the University of Edinburgh. Thank you, Grace Garland, Pooja Kishinani, Sherilyn MacGregor, David Schlosberg (again!) and Heather Urquhart. The final product is infinitely better for your insights.

Finally, thank you to my lovely parents and in-laws, Vivien and Harry, Kate and David, for endless encouragement (and hours of patient childcare). To Tom and Sarah, for believing in me (and putting up with me!). To my little girls, who keep me going, even when the writing – and what I'm writing about – gets desperately hard.

Elizabeth Cripps
Edinburgh
November 2021

INDEX

INDEX

INDEX

INDEX